AF615443

HAILSTORMS AND HAILSTONE GROWTH

suggestions on various parts of the book. I am happy to extend my thanks to Miss Sally Young for her invaluable assistance in preparing the manuscript.

The author is also greatly indebted to many societies and the publishing houses for permission to reproduce many of the diagrams; these contributions have been acknowledged at appropriate places in the text.

Albany, N.Y.
November 1975

N.R.G.

CONTENTS

HAILSTORMS

AND

HAILSTONE GROWTH

Narayan R. Gokhale
Professor of Atmospheric Science
State University of New York at Albany

State University of New York Press
Albany, 1975

Hailstorms and Hailstone Growth
First Edition
Published by State University of New York Press
99 Washington Avenue, Albany, New York 12210

Printed in U.S.A.

Library of Congress Cataloging in Publication Data

Gokhale, Narayan R. 1924-
Hailstorms and hailstone growth.

Bibliography: p.
Includes index.
1. Hail. I. Title.
QC929.H15G64 551.5'787 75-19480
ISBN 0-87395-312-6
ISBN 0-87395-313-4 (microfiche)

PREFACE

Hailstorms and hailstone growth is a subject about which our knowledge and understanding is expanding at an impressive rate, largely because it now seems possible that a measure of control may be exercised over hailfalls. But if this intervention is to be effective, it must be based upon a sound understanding of the physical processes involved in the growth of hailstones. It is in an attempt to summarize and place in perspective the developments which have taken place in this field in the last two decades that this book has been written.

Our new knowledge would not have been obtained without the efforts of a few dedicated groups who have pursued the subject in different parts of the world. In giving an up-to-date treatment of the studies of hailstorms and hailstone growth, the present author salutes the work which has been carried out to a common end in many countries.

Within the last decade there have been a few monographs dealing with some aspects of the subject, but none is as comprehensive as the present work aspires to be. This book, the first of its kind, is a reference manual, gathering in a single source a comprehensive and detailed account of modern researches on hailstones and hailstorms. It presents an extensive review of the field. The morphology of hailstones, the internal and external structures of hail, the spatial and temporal characteristics of hailstorms and hailfalls, the radar characteristics, forecasting, and suppression of hailstorms are discussed in such a manner that facts and statistics are presented which are not found elsewhere. The text includes an integrated account of the subject and major reference works to the end of 1975.

The book has been written with two main purposes in mind. In the first place it is designed to serve as a textbook for advanced graduate courses in the subject. It is intended for the serious student of physical sciences and technology, and presupposes that the reader has taken courses in general meteorology and cloud and precipitation physics. It is designed to give the student the necessary background to carry out independent work. Secondly, by including copious references to recent original papers it is hoped that it will provide a useful reference volume for research workers in the field of hailstorms and hailstone growth and help focus the attention of the scientific community on the most important and urgent problems in hail research.

I am indebted to Dr. A. E. Carte of the National Physical Research Laboratory, Pretoria, South Africa, Dr. G. Goyer of the Research Council of Alberta, Canada, Professor R. G. Soulage of the University of Clermond-Ferrand, France, and my colleagues at the State University of New York at Albany, Drs. H. Hamilton and J. E. Jiusto, who have read the manuscript and offered helpful

PAGES

PAGES

PAGES

PAGES

PAGES

CHAPTER 1

INTRODUCTION

Thunderstorms are violent and spectacular phenomena of atmospheric convection, always associated with heavy rain, gusty winds, and lightning, often with hail, and occasionally with tornadoes. Although medium or large hail-producing storms are usually associated with thunderstorms, small hail may fall from clouds which do not reach thunderstorm proportions. On the other hand, many thunderstorms may contain small hail at some stage of their development, but the embryos may be in localized regions, and they may melt before reaching the ground. Therefore, the relationship bewteen the occurrence of thunderstorms and hailstorms is rather difficult to determine precisely.

Research into the origin and subsequent control of hailstorms is a field which until 1950 attracted little attention among atmospheric scientists. No systematic effort had been made to investigate hailstorms, though one heard occasional reports of unusual sizes or shapes of hailstones causing loss of crops, breakage of windows, damage to roofs, and denting of cars. Sometimes large hailstones were reported to be lethal, killing or injuring birds and animals.

Increased knowledge of the formation of hail has worldwide significance since hailstorms occur all over the globe. It is known that such storms usually develop in the afternoon or evening during the summer season. Frequently, these storms occur in the continental interiors of middle latitudes and their occurrence diminishes toward the poles and over the sea. It should be emphasized, however, that there are few measurements in these latter regions.

A more comprehensive study of this phenomenon is appropriate because of man's need to understand and control hailstorms. Hail suppression has economic significance, since hail does a great deal of damage every year in many countries. In the United States the annual damage to crops and property from hail is estimated at some $300 million, which in a typical year averages somewhat more than the loss due to tornadoes. It is likely that more than 80 percent of the total value of the hail damage is due to crop loss. This percentage depends on many factors besides the frequency of hailstorms, factors such as the percentage under cultivation of the total affected land, the monetary value of the crop, and the vulnerability of the crop to hailfalls. Although severe hailstorms occur in Colorado and Wyoming, the total amount of damage is much lower in these states than in Kansas, where the hailstorms are less severe but come during the period when the wheat crop is maturing. The frequency of hailfalls in North Carolina is very low, but the damage is considerable to the expensive and vulnerable tobacco crop. It is an unpleasant coincidence that intense hailstorms occur in the United States during the time when crops are most vulnerable: July in the mid-western corn belt and August in the northern wheat country. The hail insurance premiums on crops cost up to 20 percent of the total value of the crops in the United States. Some success towards mitigating hailstorms would therefore be expected to repay handsomely any funds expended in hail research.

Hail is one of the most serious hazards faced by pilots, since aircraft hit by hail can be severely damaged by exposure of the light metal, especially if the aircraft is traveling at high speed. Of course, with the help of airborne radar, pilots now try to avoid even entering severe thunderstorms. But the chances still remain for an unexpected and costly encounter with hail and severe turbulence.

New research capabilities have increased hailstorm studies in recent years. At times, an isolated hailstorm not embedded in a large frontal system can easily be tracked by radar, research aircraft, and other instruments. Some aspects of its structure can be studied throughout its period of development, maturity, and dissipation, which usually does not extend more than a few hours. In additon, the knowledge of hailstorms gained over the years, as well as some success in developing triggering techniques for the disruption of the process of hailstone growth, have provided great impetus to research.

Thus, the observation, understanding, prediction and, perhaps, eventual modification of these storms are being investigated by many researchers. Hail damage in the United States is a national problem, and official recognition of this fact has finally been granted. Recently many agencies of the federal government put their support behind a National Hail Research Experiment (NHRE) to study hailstorms and to investigate the possibility of hail suppression.

A study of hail formation consists of identifying the atmospheric conditions that produce hailstorms and the cloud micro-physical processes of hailstone growth. So far it has not been possible to make a complete model for hailstorms which incorporates air motion and particle spectrum in space and time from their initiation to their eventual dissolution and precipitation, respectively. However, we can attempt to synthesize carefully the many pieces of information that have been obtained on hailstorms by many investigators over the globe in hopes that a more unified picture will emerge; such is the goal of the succeeding chapters. They present a current review of the field including the morphology of hailstones, the spatial and temporal characteristics of hailstorms and hailfalls, the internal and external structures of hailstones, models of hailstone growth, the radar characteristics, and the forecasting and suppression of hailstorms.

CHAPTER 2

CLIMATOLOGY OF HAIL

As mentioned in the introduction, hailstorms occur frequently in the continental interiors of middle latitudes. There are fewer over land in tropical and polar regions and fewer generally over the sea. In the tropics, the development of vigorous and well developed cumulonimbus is rather rare and the melting in the subcloud layer is maximum. In cold climates, the moisture as well as the vigorous development of the cloud are limited. Over the oceans, the lack of intense surface heating inhibits the development of strong updrafts to produce large hailstones. Thus, the climatology of the high latitudes and the oceans should produce hail only infrequently, but there is little data concerning its occurrence there. This distribution pattern has considerable significance when one realizes that the people and crops are concentrated in the middle latitudes that are vulnerable to hailstorms. The geographical regions most liable to hail damage include the United States in the area of the Great Plains east of the Rocky Mountains, Canada, Europe, Central and Northern India, the Argentine Pampas, Kenya, South Africa, the USSR and China.

2.1 Hailstorms in the United States:

Hailstorms occur in nearly all sections of the United States according to the records of the National Weather Service.

The rate of occurrence and crop damage, however, is much greater in certain states than in others. A detailed distribution according to area is discussed by Flora (1956). The state of Kansas leads the nation in the total amount of damage to crops from hail and is followed by several other states in the general area between the Rockies and the Mississippi River. There have been attempts to correlate the crop damage and severity of the storm. However, such statistics would be of limited value because damage depends upon the type of crop affected and the time in the life cycle of the crop at which the storm occurs.

2.1.1 Hail-thunderstorm ratio:

There are well-known geographical and seasonal variations in the hail-thunderstorm ratio. Forty years of data (U.S.W.B. Hydromet. Rep. No. 5, 1947) indicate a maximum annual thunderstorm frequency in the Gulf States; however, this region and Atlantic coastal areas show hail minima. The hail frequency increases toward the west, attaining a maximum in the Great Plains just east of the Rockies. There is a secondary maximum along the Pacific coast; however, only a few storms are involved, and these produce small ice pellets instead of medium or large size hail.

The monthly variation of thunderstorm and hail frequencies and the hail-thunderstorm ratio for the entire United States over a period of forty years are shown in Fig. 2-1. The seasonal variation in the hail-thunderstorm ratio is minimal in the coastal

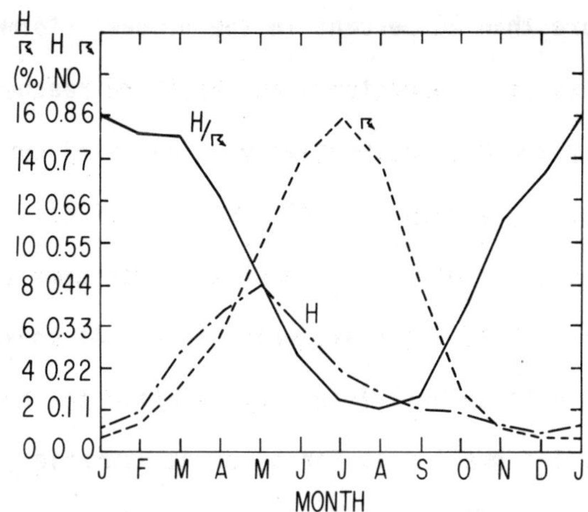

FIG. 2-1 Mean monthly values of thunderstorm (R) frequency, and percentage of thunderstorms resulting in hail (H/R) in the United States [1904-1943]. [Data from U.S.W.B. Hydrometeorological Report No. 5, 1947]. (from Appleman, 1959)

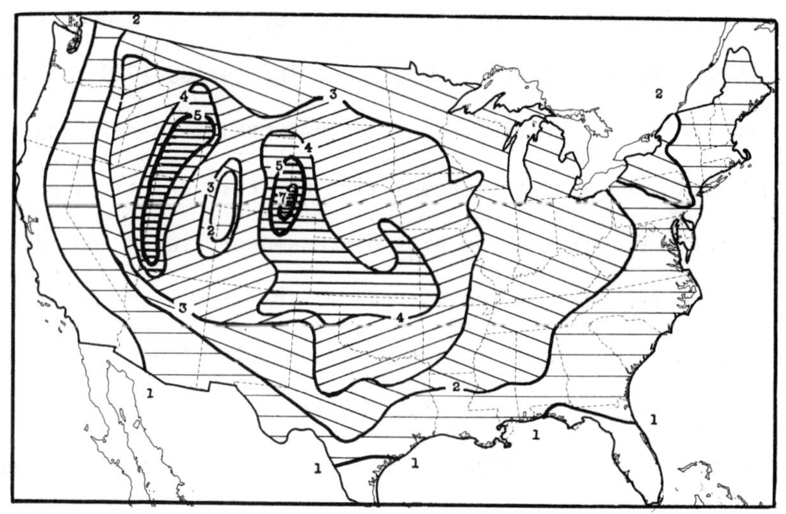

FIG. 2-2 Normal annual number of days with hail at a point. (from Visher, 1966)

storms and more than 20 percent in the midwest storms. The maximum frequencies of thunderstorms and hail are reached in the months of July and May, respectively. The third curve in Fig. 2-1 shows, interestingly enough, that the ratio of hail to thunderstorms reaches the minimum at almost the same time the thunderstorm frequency is at its peak, which is as yet unexplained.

The data on hail frequency relative to thunderstorms indicate useful information as far as point summaries are concerned. However, when area figures are summarized, definite bias is introduced if thunderstorm day counts are reported from a single station. In addition, a truer picture of the incidence of hail for a given area is depicted by network reporting than by point reporting. (Fig. 2-2).

Analysis of hailstorm frequency for the Denver area has been reported by Beckwith (1960) over the period 1949 to 1958 with the help of a cooperative reporting network. The hail-thunderstorm ratio shown in the second line of Table 2-1(a) is a useful indicator. But it has the bias discussed earlier. The ratio of 1:8.0 in a ten year period is reduced to 1:1.7 when such bias is not present. The area frequency versus point frequency of hail in the Denver area is shown in Table 2-1(b). The official hail days in column 3 are those reported by the Weather Bureau. Ten year average of the area to point frequency ratios is 4.4:1.0.

TABLE 2-1(a)

Hail-thunderstorm ratio.

(from Beckwith, 1960)

Reporting	Total hail days	Total thunderstorm days	Ratio
10-year unofficial area	225	391	1:1.7
10-year official point	51	391	1:8
59-year official point	4*	43*	1:11

*These are annual averages.

TABLE 2-1(b)

Area frequency versus point frequency of hail.

(from Beckwith, 1960)

Year	Network hail days	Official hail days	Area to point ratio
1949	33	4	8:1
1950	23	3	8:1
1951	34	9	4:1
1952	16	2	8:1
1953	21	7	3:1
1954	9	4	2:1
1955	26	9	3:1
1956	13	3	4:1
1957	25	6	4:1
1958	25	4	6:1
Totals	225	51	4.4:1 *

* Ten year average.

2.1.2 Annual and diurnal variation:

Hailstorms are relatively rare from September to early April and account for only 15 percent of the total number of hail occurrences. The annual variation is shown in Fig. 2-3. In many states the active season is May, June, July and August; these months account for 85 percent of the hailstorms. The variation in frequency in the Central United States is discussed in detail in Section 2.1.5.

The diurnal variation of hailstorms is shown in Fig. 2-4. There is a sharp rise between noon and 10 P.M., so that this period accounts for 85 percent of the total number of storms. The histogram indicates that the frequency maximum is between 4 and 6 P.M. This trend is confirmed for storms in Northeastern Colorado by Goyer and Wood (1972).

2.1.3 Economic losses from hailstorms:

The economic losses produced by various forms of severe weather in the United States are difficult to ascertain, and hence very little information is available on the subject. However, crop losses due to hail are better defined because of well kept records by the insurance companies over a period of more than 40 years. It should be realized, though, that the total damage due to hail can only be approximately estimated using the insurance data since 80 percent of the crop value is not insured.

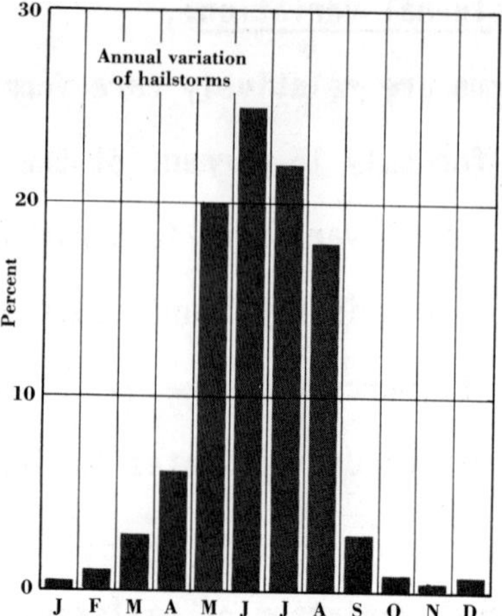

FIG. 2-3 Annual variation of number of hailstorms in the United States. Storms are rare from September to early April. (from Petterssen, 1969)

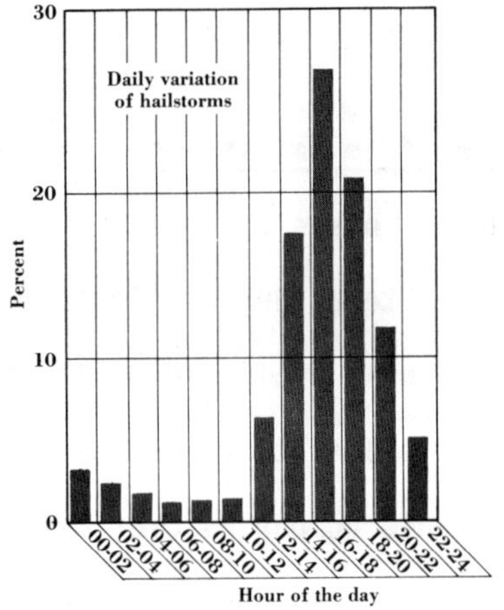

Fig. 2-4 Diurnal variation of hailstorms in the United States. About 85 percent of the storms occur between noon and 10 P.M. (from Petterssen, 1969)

Most crops grown in the United States are susceptible to hail damage, and some, such as fruits and tobacco, are more easily damaged than others. The major crops and their percentage losses as reported by Changnon (1971) (as a percentage of the 1963-1967 national estimated total loss) are wheat - 51%; cotton - 11%; corn - 10%; soybeans - 9%; and tobacco - 7%.

The annual average crop loss due to hailstorms is estimated by Jones (1969) to be $284.1 million (at 1968 price levels). The regional crop losses in the United States are shown in Fig. 2-5. The Great Plains area ranks first nationally for highest crop damage and the Corn Belt region ranks second. Thus the large losses are in the states called the 'breadbasket' of the nation.

Property loss involves primarily livestock, trees, vehicles and structures. Data are unavailable on a national scale. However, two detailed studies in Illinois indicate a crop-property loss ratio of 9:1. Using the figure $284.1 million for crop-loss, the annual estimated average property loss from hail is $31.6 million. Thus the total loss per annum due to hail damage in the United States is about $315 million.

2.1.4 Diversity of climatology:

The climatology of hail in the USA is very diverse and has many facets. The diversity can be seen from the map shown in Fig. 2-6. Hail damage is restricted mainly to the months of April, May, June, July and August. Figure 2-7 shows the month

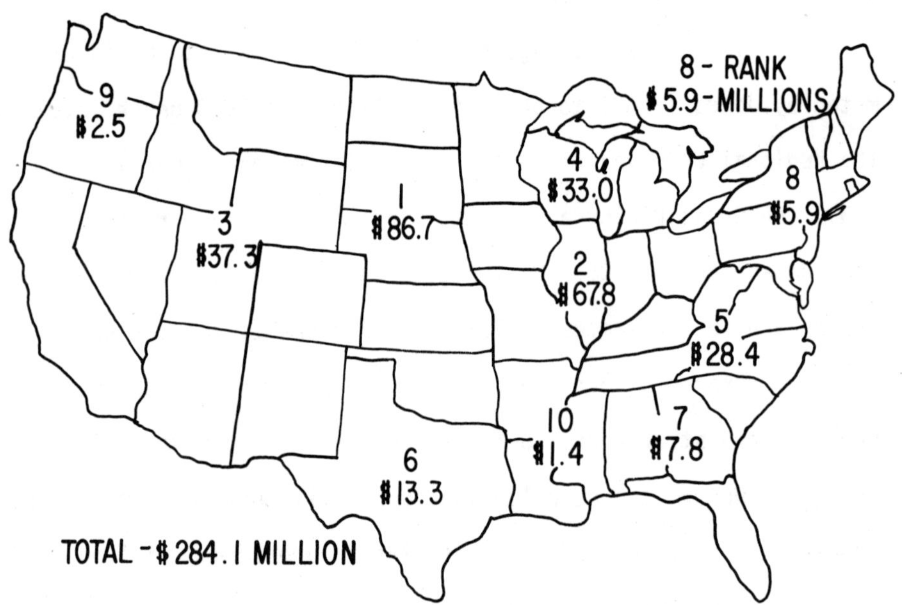

FIG. 2-5 Regional average annual crop-hail losses [1948-67] and their rank. (from Changnon, 1971)

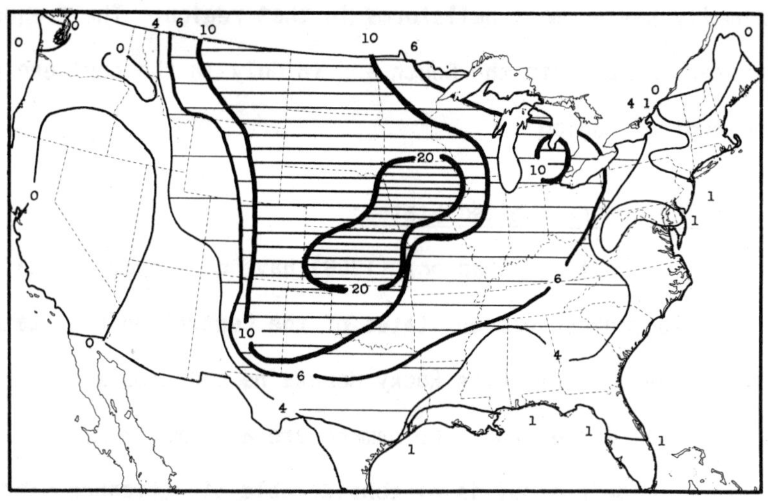

FIG. 2-6 Normal number of hailstorms per growing season [based on state averages]. (from Visher, 1966)

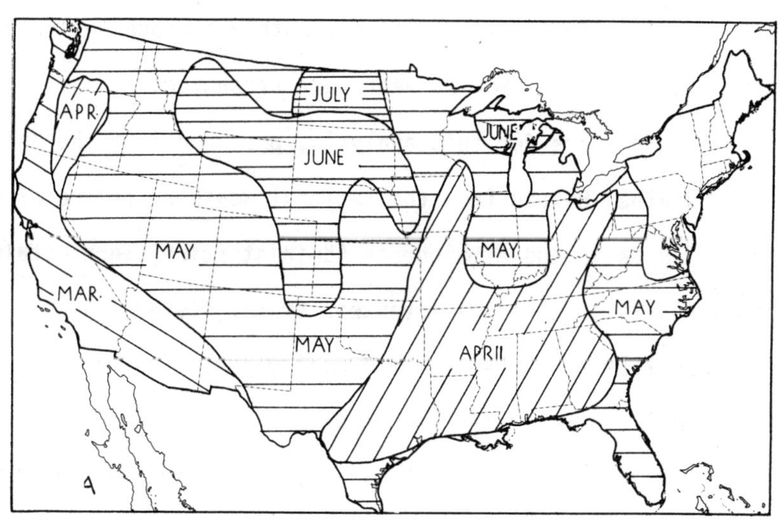

FIG. 2-7 Hail: the month normally having most hailstorms. (from Visher, 1966)

normally having most hailstorms in that region. The frequency is greater in April in the South and in July in the northern part of the central region.

2.1.5 The Central United States:

Stout and Changnon (1968) have developed a detailed hail climatology of seventeen states of the Central United States. In these areas, between the Rocky Mountains and Appalachian Mountains, hail losses and crop damage are at a maximum.

Their study is of considerable significance since it is based on a much larger amount of data than earlier studies published by the Weather Bureau. Their atlas includes 19 maps of hail patterns which provide information on the spatial as well as temporal variations. In addition, monthly average hail-day patterns and the totals for 20 years are presented.

The data for this analysis, procured from the United States Weather Bureau (now the National Weather Service of the National Oceanic and Atmospheric Administration) were obtained from 1285 stations in the seventeen states listed in Table 2-2. These observations were recorded mostly during the period 1910 to 1960. Most of the states had between 2 and 5 first-order stations, and the number of such stations scattered through the seventeen states was 87. These stations reported reliable information as they were manned by trained meteorologist. The point average of the number of hail days for each month is shown in

TABLE 2-2

Hail data stations.

(from Stout and Changnon, 1968)

State	Number of Stations		Total	Square Miles per Station*
	10-to19-yr Records	Greater than 19-yr Records		
Illinois	28	61	89	634
Indiana	8	31	39	930
Michigan	18	52	70	832
Wisconsin	12	37	49	1145
Missouri	59	40	99	703
Iowa	26	41	67	840
Minnesota	40	38	78	1077
No. Dakota	53	36	89	794
So. Dakota	37	46	83	928
Nebraska	45	48	93	830
Kansas	52	59	111	741
Oklahoma	19	32	51	1371
Texas	62	28	90	2970
New Mexico	22	27	49	2483
Colorado	16	38	54	1930
Wyoming	26	34	60	1632
Montana	77	37	114	1290
Totals	600	685	1285	21130

* $1\ mi^2 \simeq 2.56\ km^2$

Table 2-3. The numbers of maximum hail days at certain locations are much higher as shown. The months September to January have the lowest hail frequency whereas April, May and June have the maximum hail frequency.

2.2 Climatology of hail in Alberta, Canada:

As mentioned earlier, a real indication of the hail patterns in a given region can be obtained from the reports of a dense network of voluntary observers. A synoptic network of standard weather observing stations is not adequate for the purpose as the number of stations is limited and they are too scattered.

The Alberta Hail Studies Project (ALHAS) has received over a period of ten years (1957-1966) more than 30,000 hail reports from a network of 25,000 farmers in Central Alberta. A summary by months of various hail parameters is shown in Table 2-4. The main hail months are June, July and August. The probability of hail occurring at some point in the area is greatest in June and July. Large hail of diameter greater than 2.5 cm reaches its maximum frequency in August. The season is effectively over by mid-September. The hail frequency varies between 1 and 6 in different areas of the province.

The relationship between physiography and the hail distribution pattern has been reported by Powell (1961). Hail appears to occur more frequently on lee slopes and in fairly flat

TABLE 2-3

Point averages of number of hail days in twenty years for different months.

MONTH	Average number of hail days in 20 year period	Maximum number of days per year in 20 years at certain locations	
JANUARY	<1	4	Southern Missouri
FEBRUARY	1	1-7	Western Colorado, NW New Mexico
		>6	Ozark plateau
MARCH	1	17	Texas, Oklahoma, Kansas, Missouri
APRIL	8	20	Texas, Oklahoma, Kansas, Missouri
MAY	10	40	Cheyenne
		25	Denver
JUNE	9	>15	Mountain hail areas of Montana, Wyoming, Dakota, Denver
JULY	1	36	SE Wyoming, Cheyenne
		10	Minnesota, Wisconsin, Michigan, N. Dakota, S. Dakota, Iowa
AUGUST	1	25	Cheyenne, Los Alamos (New Mexico)
SEPTEMBER	<1	15	upper peninsula Michigan
OCTOBER	<1	20	Michigan
		12	Colorado, New Mexico
NOVEMBER	<1	20	region around Great Lakes
DECEMBER	<<1	4	South Central New Mexico

TABLE 2-4

Summary by months of various hail parameters, in Alberta, Canada.
(from Summers and Paul, 1967)

Hail Parameter	May	June	July	Aug.	Sept.	Total o average
Average number of hail days	5.3	20.3	20.3	15.7	4.9	66.5
Probability of a hail day	17%	68%	66%	51%	16%	48%
Average no. major hail days	0.2	5.4	5.4	2.3	0.2	13.5
Prob. of a major hail day	1%	18%	17%	7%	1%	10%
Total no. mailed-in reports	484	8968	9672	4725	378	24227
Mean hail onset time (MST)	1527	1607	1655	1612	1517	
Mean hailfall duration (min.)	11	9	11	11	8	10
% reports with hail > 1" diam.	8%	11%	22%	23%	1%	18%
% reports with soft hail (1965-66)	52%	59%	35%	42%	59%	44%

valleys, and less frequently around lakes and crests of hills or ridges. These findings are in direct contradiction to the observations analyzed by Stout and Changnon (1968) in the Central United States.

Hail damage in Alberta constitutes a serious problem for agriculture (Summers and Wojtiw, 1971). During one hail season, claims for crop damages in some areas within the province were filed on 50 separate days. The estimated direct losses of field crops are in the neighborhood of $23 million annually. In addition, there are secondary losses in the affected areas, mainly due to decrease in general business activity. Occasionally a severe storm hits one of the larger urban areas. The resultant property damage, mainly to structures and vehicles, can then run into several millions of dollars. The total economic annual loss due to hail damage is estimated to be between $30 and $40 million.

The correlation coefficients between hail damage and various hail parameters are shown in Table 2-5. The hailstone size and the total mass of hail are the two parameters which correlate highly with hail damage.

2.3 Climatology of hail in Europe:

The major hailfalls in Western Europe over a period of three years, 1968 to 1970, are shown in Figs. 2-8 to 2-10. The areas are ascertained from the insurance data on yearly hail damage. The areas of intense hailfall vary from year to year. The countries having frequent hailfalls are France, West Germany, Austria, Northern Italy, Denmark, Switzerland, Greece and the southern part of Sweden. May, June, July and August are the months during which most of the hailfall occurs.

The Po Valley of Northern Italy (Fig. 2-8) has one of the world's worst crop-hail problems (Morgan 1973). The high value and the sensitivity to hail damage of the crops, mostly grapes and tree fruits, yield a very high loss per hail event, the estimated average loss being $1330 per square mile per hail season, nationwide, with an average loss reaching $7100 per square mile on the smaller scale in the Po Valley during April-September hail season.

TABLE 2-5

Correlation coefficients between hail damage and various hailfall parameters, in Alberta, Canada. (from Summers and Wojtiw, 1971)

Parameter	Correlation Coefficient with Damage
Maximum hail size	0.52
Modal hail size	0.50
Mass per unit area	0.49
Impact momentum	0.47
Impact energy	0.40
Softness	-0.19
No. of days after June 1	0.16
No. of stones per unit area	0.002

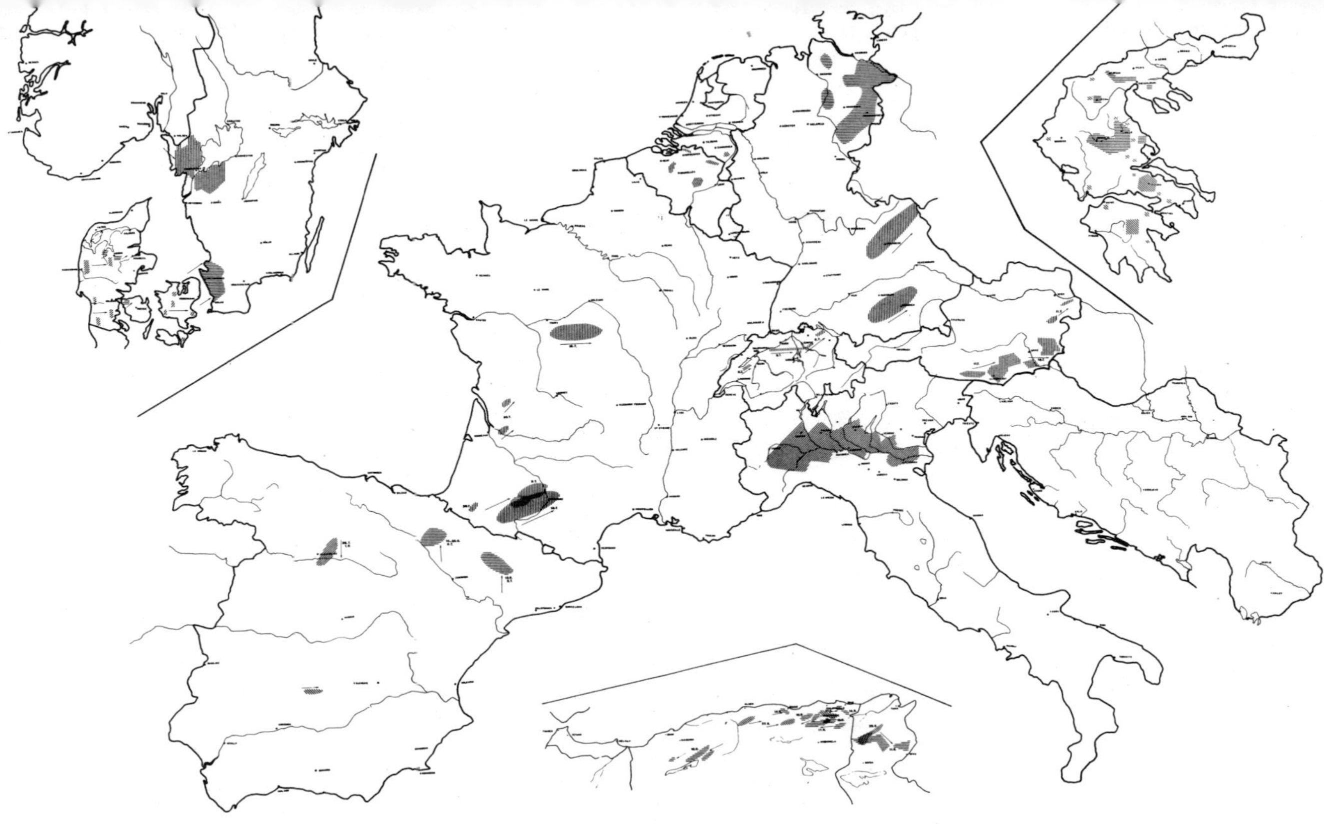

FIG. 2- 8 The major hailfalls in Europe during 1968. (Source: Swiss Hail Insurance Company, Zurich)

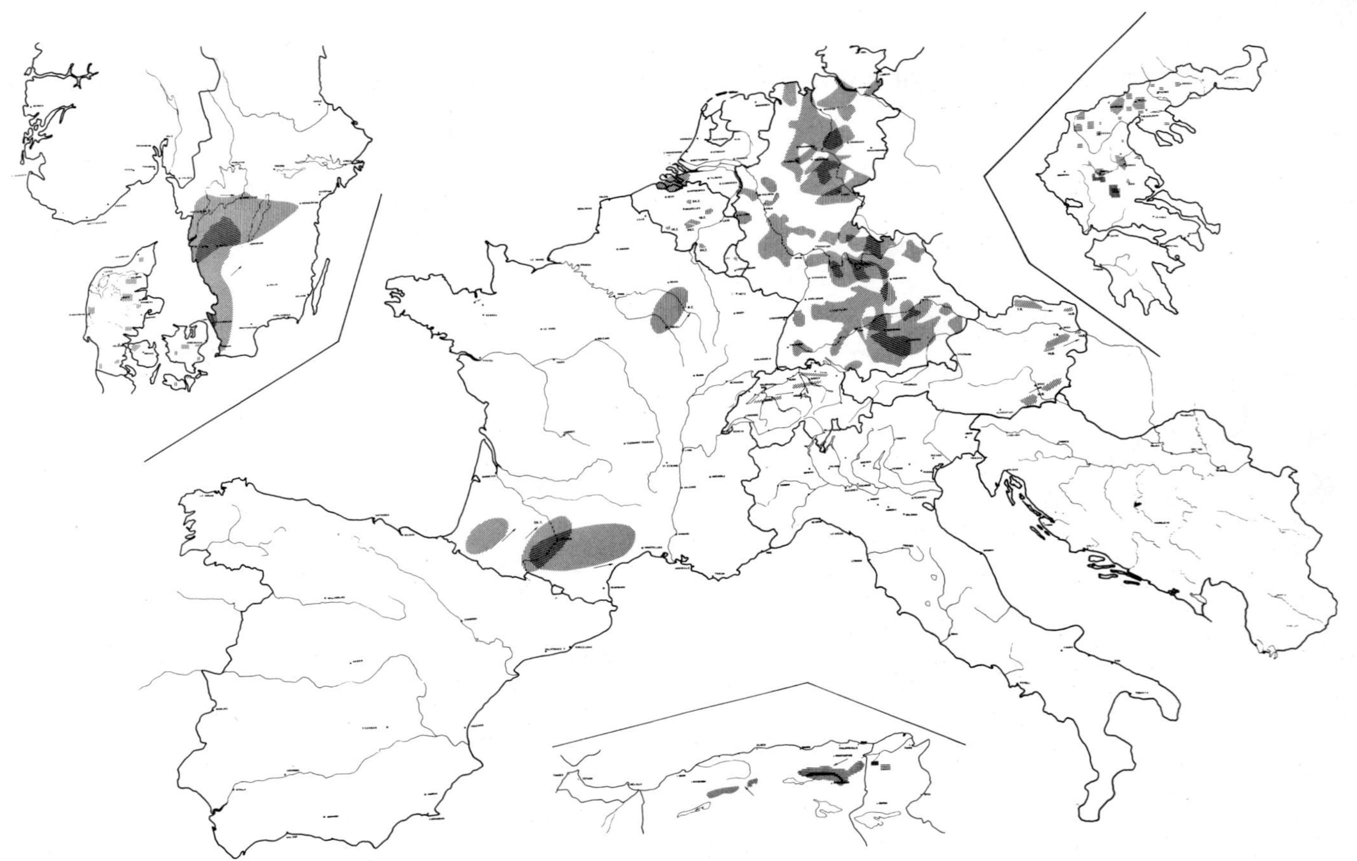

FIG. 2- 9   The major hailfalls in Europe during 1969. (Source: Swiss Hail Insurance Company, Zurich)

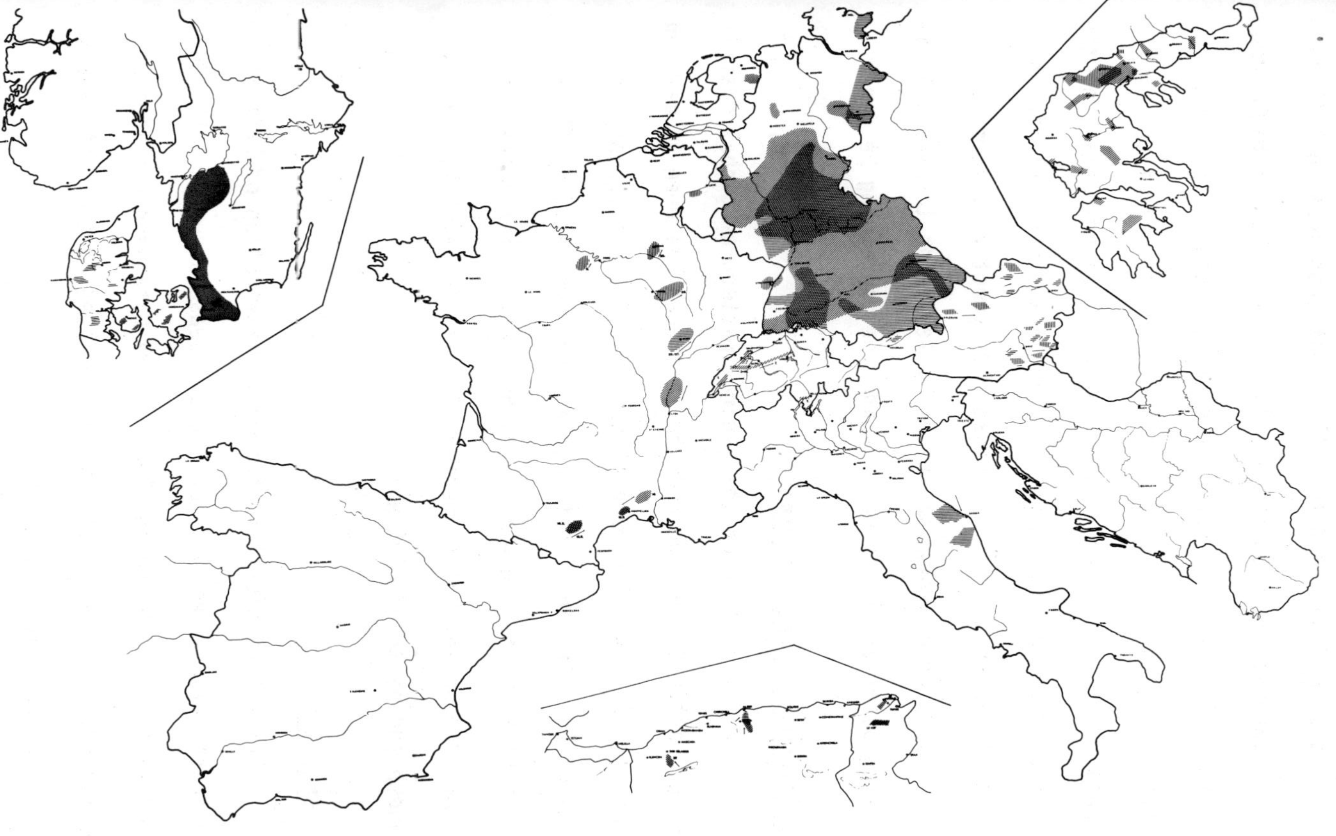

FIG. 2-10 The major hailfalls in Europe during 1970. (Source: Swiss Hail Insurance Company, Zurich)

It should be noted that the great crop loss to hail is not due to excessively frequent yearly hailfall, but to high crop value, high hail frequency during the growth season, frequent large hail, long hailfall duration, and high concentrations of hailstones per square meter. The meteorological uniqueness of the region is ascribed by Morgan to the terrain and the cyclonic development it causes during the passage of synoptic systems.

Heavy crop damage in the Po Valley tends to have political repercussions as the farmers demand that something be done.

2.4 Hailstorms in other countries:

The information on the general pattern of global distributions of hailfalls in high and low latitudes has been summarized by Lemons (1942). Also, a general summary of the frequency and severity of hail in many countries has been presented by Flora (1956). These studies show that the hailfall is significant in Argentina, the Soviet Union, Mexico, Brazil, India, Pakistan, Bangladesh and parts of Africa.

2.4.1 Argentina:

The geographical and temporal distribution of hail in the Northern part of Mendoza for the six seasons during 1957-1963 has been reported by Grandoso (1966). The data were provided by 15,000 farmers from the total damage estimated to their vineyards which are spread over an area 1200 km^2.

The number of hail days per season varied from 2 to 10 as shown in Fig. 2-11. This is not surprising because hailfall is such a localized phenomenon; its observation depends markedly on the density of the reporting network.

The seasonal variation showed the maximum frequency of hail occurrence and the maximum total damage to be in January. December was in second place, and the minimum was reached in October for both variables. However, there was a large variability in monthly hail damage from year to year because of the occurrence of a few severe hailstorms.

The diurnal variation of total damage indicated a peak late in the afternoon or early evening as shown in Fig. 2-12. In fact, 90 percent of the damage occurred between 1600 and 0200, out of which 36 percent occurred from 1900 to 2100.

A few other interesting observations are worth noting. The geographic differences in total damage were statistically significant for an 8 percent significance level. Both the hail frequency and the total damage increased with distance away from the mountains, the limit being 100 km.

2.4.2 Soviet Union:

In the Soviet Union, hail occurs in Armenia, Georgia, Azerbaidzhan, Moldavia, Turkmenistan, the Ukraine, Transcaucasia and the Northern Caucasus. The maximum frequency of hailfall, about 8 to 10 days annually, occurs in Transcaucasia, the

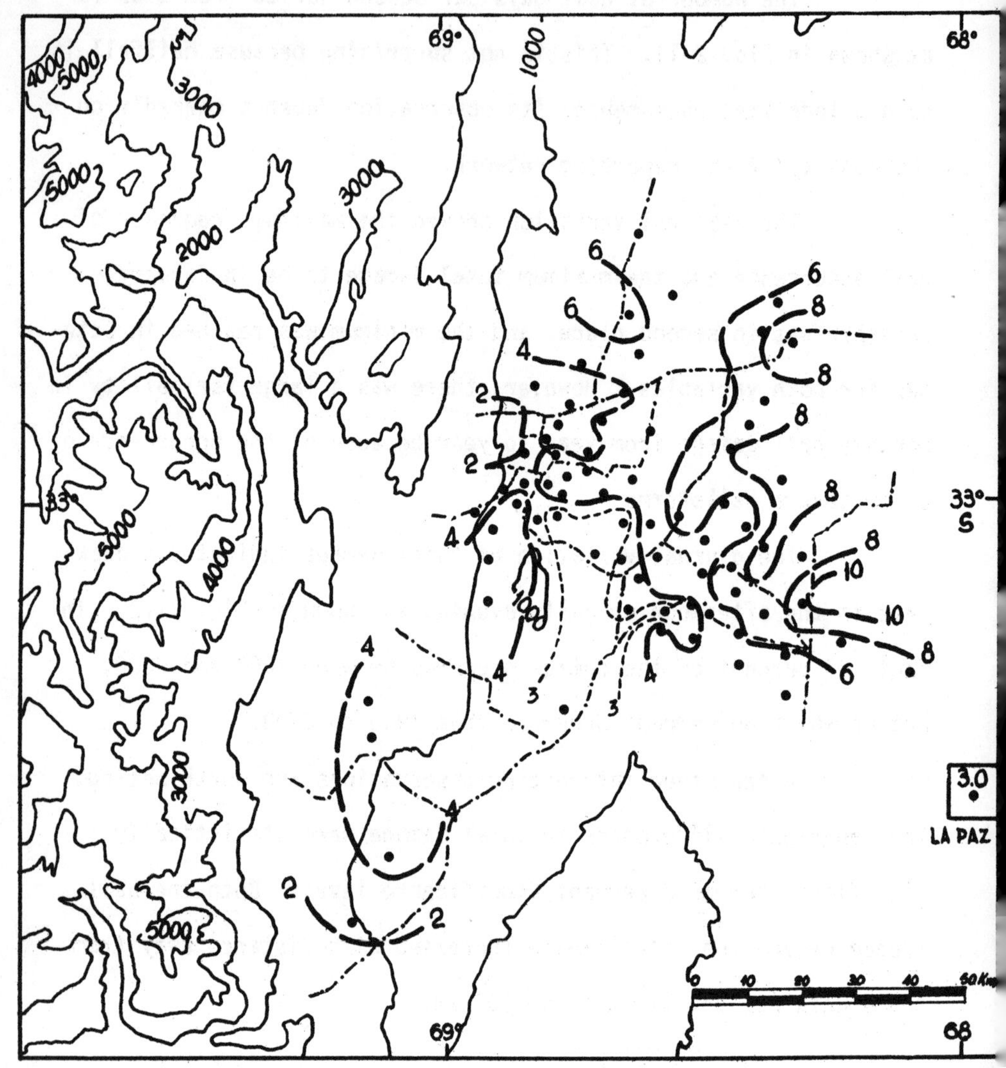

FIG. 2-11 Geographical distribution of the number of hail days per season, [1957-1963]. Points show the centers of 72 geographical units. (from Grandoso, 1966)

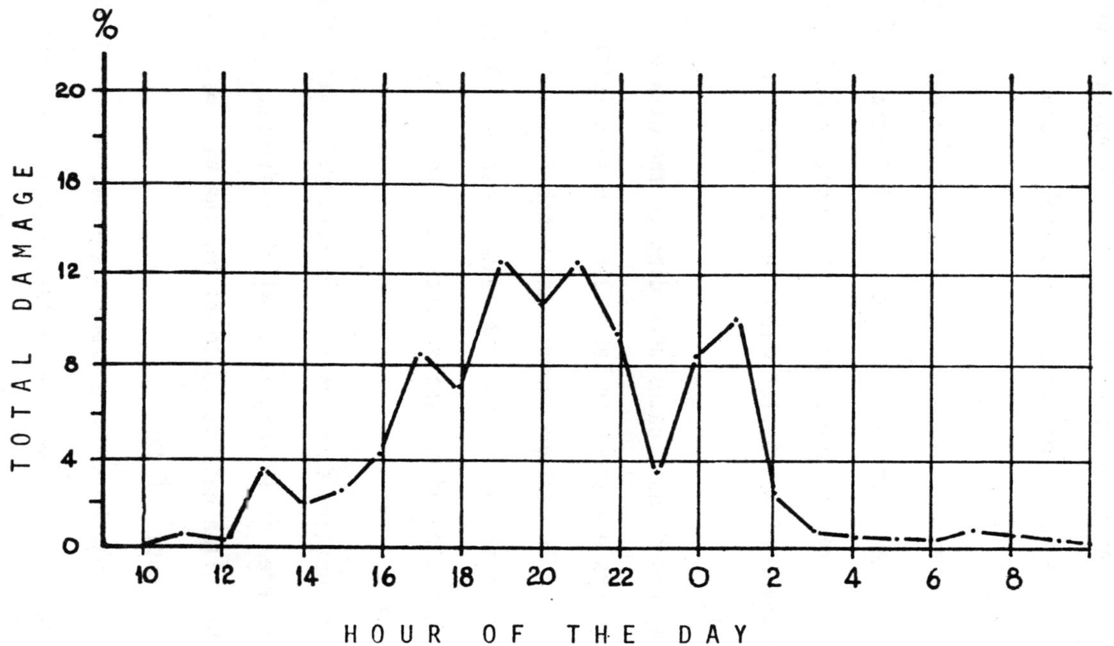

FIG. 2-12 Diurnal distribution of total damage due to hail [1957-1963].

(from Grandoso, 1966)

Northern Caucasus and in the south of Kazakhstan. The average number of hail days at given locations in the USSR appears not to exceed four. In the region of Moscow, it is only 2 days per year (Sulakvelidze, 1967).

The observations of recent years have shown considerable hail damage in Turkmenistan and other regions of Central Asia. Intense hail damage has occurred around Leningrad, Yakutsk and other northern areas of the Soviet Union, extending up to the 60-65th parallel.

Dimensions of hail paths according to the data of different investigators are shown in Table 2-6, and these differ in different regions. In western Georgia the width is on the average 10 km, whereas in eastern Georgia, the length reaches 100 km. Hail path widths are based on limited data, and hence the figures may not be reliable.

In the Northern Caucasus and Transcaucasia, hail damage, if it exceeds an area of $100km^2$, is usually disastrous. When the thickness of the fallen hail is 20 to 30 cm, the hail may lie on the ground for a couple of days.

In the USSR, Poland and Czechoslovakia, hail falls most often during the spring, summer and autumn months. The maximum hailfall observed in the period 1938-1959, was in May and June; none was observed from November through March. Table 2-7 gives the average number of hail days per month in various regions.

TABLE 2-6

Dimension of hail paths according to the data

of different authors.

(from Sulakvelidze, 1967)

Author	Location	Length (km)		Width (km)	
		maximum	mean	maximum	mean
Prohaska	Austria...........	several 10^2km	10-20	10-20	8-9
Schaefer	Grand Island......	48	-	8	-
Gigineishvili	Eastern Georgia...	90-100	20-30	10	5-7
Lemons	USA...............	-	-	-	1-6
Chepovskaya	Northern Caucasus.	400	15-20	8-10	1-6
	Mean..............	-	15-20	-	4-7

In 80 to 90 percent of the cases, the diameter of hail falling in USSR does not exceed 2 cm. Hailstones of diameter 5-7 cm are a rare phenomenon.

2.4.3 Hail incidence in the tropics:

Frisby and Sansom (1967) have compiled data on the incidence and frequency of hailfall in the tropical areas, within the latitudes 23 1/2 degrees north and south of the equator. These data were gathered over a number of years from literary sources and by correspondence with individuals as well as government meteorological services. The resulting document makes a comprehensive assessment of the tropical hail phenomenon. In general, the hail problem is definitely less serious in the tropics than at mid-latitudes: both hail frequency and hail size are of

TABLE 2-7

Average number of days per month with hail for various regions.

(from Sulakvelidze, 1967)

MONTH	Eastern Georgia (according to Gigineishvili and Bartishvili)		Poland (according to Kozminski)		Yugoslavia (according to Labovich)		South of France (according to Genève)	
	average number of days	%	average number of days	%	average number of days	%	average number of days	%
Jan	-	-	-	-	0.3	13	0.7	12
Feb	-	-	-	-	0.3	13	0.3	5
March	-	-	-	-	0.2	9	0.8	14
Apr	0.8	13	0.5	2	0.1	4	1.0	18
May	1.8	29	5.4	19	0.1	4	0.8	14
June	1.5	25	5.1	17	0.1	4	0.3	5
July	0.7	11	9.0	32	0.2	9	0.2	4
Aug	0.6	10	6.2	22	0.1	4	0.1	2
Sep	0.5	8	1.3	7	0.1	4	0.2	4
Oct	0.2	3	0.3	1	0.2	9	0.8	14
Nov	-	-	-	-	0.2	9	0.2	4
Dec	-	-	-	-	0.4	18	0.2	4
The whole year	6.1	100	27.8*	100	2.3	100	5.6*	100

* Corrected totals are included by the author.

considerably less magnitude, except for the frequent hailfalls in Kenya. Nevertheless, when hail occurs, it causes major damage to expensive crops such as tobacco, tea and coffee.

For hail incidence in the tropical areas where hail is infrequent, the reader may refer to the paper by Frisby and Sansom (1967). The hailstorm distributions in Africa, India, Pakistan and Bangladesh are discussed in the sections to follow.

2.4.4 Africa:

Though the thunderstorm frequency over certain parts of Africa is very high, hail occurrence is relatively rare in many areas. Figure 2-13 shows the mean annual point frequency of hailstorms over Africa. But these values may not be very accurate, since they are based on information obtained from many sources and cover varying periods. According to Sansom (1966), hailfall is totally absent between 20°N and 30°N and is rare on the coastline along the Atlantic and the Indian Ocean. The main hail areas are Morocco (Atlas Mountains), Rwanda, Burundi, and parts of the Congo, Western Kenya, Western Uganda, Ethiopia, Transvaal and the plateau of the Republic of South Africa, Lesotho, Swaziland and the high plateau of Madagascar. Figure 2-14 shows the months during which most of the hail occurs in any particular area.

For further details on each area, the reader is referred to the review by Sansom (1966). In addition, more information is now available on the point frequencies of hailfall in South Africa

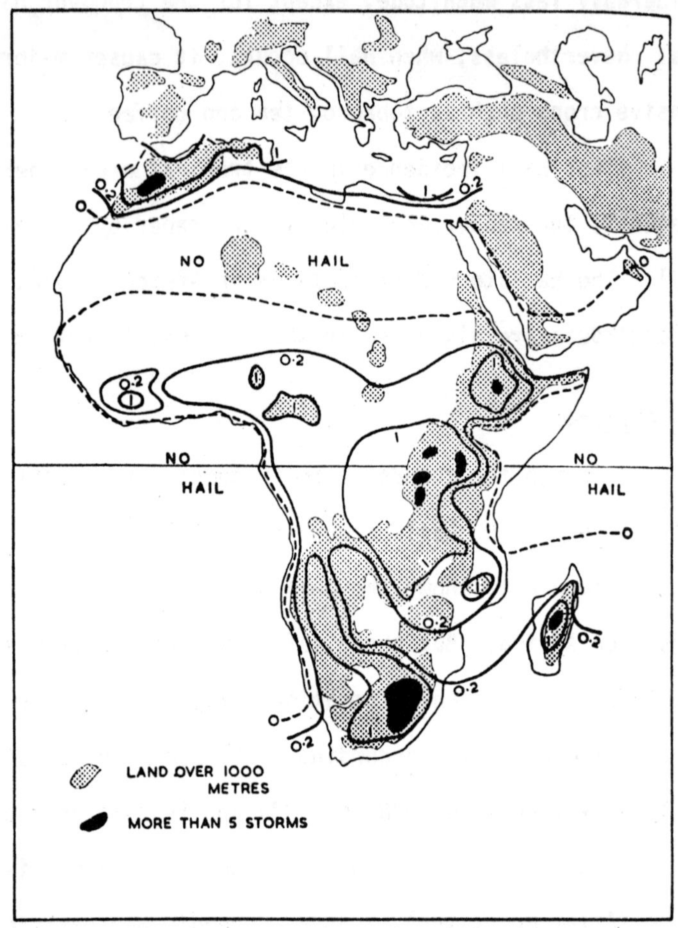

FIG. 2-13 Mean annual frequency of hail at a point.
(from Sansom, 1966)

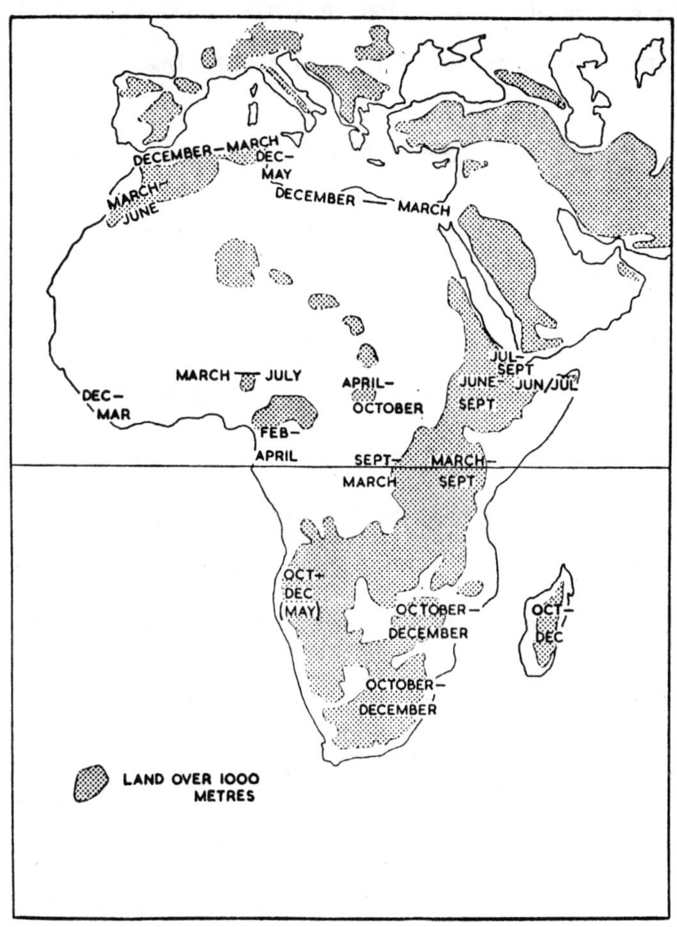

FIG. 2-14 Main hail seasons throughout Africa.
(from Sansom, 1966)

(Carte and Basson, 1970) and in Kenya (Henderson et al., 1970); this information is summarized in the following two subsections.

2.4.4.1 Pretoria-Witwatersrand region of the Transvaal, South Africa:

The hail climatology of this region has been reported by Carte and Basson (1970), for a period of seven years beginning in 1962. The data were collected by the observing network which was spread over 1000 square miles. The details of this network are described in the next chapter.

Hail fell within this area on an average of 71 days per year between 1962 and 1969. Annual averages of the number of hail days, rainfall days and thunderstorm days are given in Table 2-8. These data show good correlation on an average between rainfall days and hail days although the wettest year did not produce the greatest number of hail days. Only the point frequencies of hail occurrence are usually reported by climatological records, and these do not indicate the areal frequency and severity of the storm.

The comparison of hail frequencies at Johannesburg and Pretoria, the two big cities within the observational area, is interesting. The altitudes of the two cities are 1675 m and 1370 m above mean sea level. Johannesburg, which has a higher elevation, experienced more storms with hailstones of medium size (1 to 3 cm) than Pretoria. However, both cities experienced nearly the same

TABLE 2-8

Average annual number of hail days within the network, and average annual rainfall and number of thunderstorm days at Pretoria, 1962-1969. (from Carte and Basson, 1970)

	Number of hail days when largest hailstones were:			Rainfall in mm	Number of thunderstorm days
	Any size	>1 cm	>3 cm		
Average	71	42	3.4	652	58
Limits*: upper	95_1	64	8	1046_3	72_4
lower	51_2	24	1	386_2	31_2

* Highest and lowest yearly values (Oct. to Sept.)

1. 1964/65; 2. 1965/66; 3. 1966/67; 4. 1963/64

number of mild storms and very severe storms.

The seasonal variation of the average number of hail days within the network for the seven year period is shown in Fig. 2-15. The summer months of November, December and January are the peak months for hail of any size. The largest hailstones usually fall early in the season (October, November).

Carte (1964) describes one of the more severe hailstorms which occurred on January 15 and 16, 1964, in the middle of the night. The hail reports revealed a practically continuous path of hail 11 to 18 km wide and 64 km in length from Johannesburg to Pretoria. Very big hailstones up to 6 cm in diameter accompanied strong winds. The distribution of hailstone sizes is shown in Fig. 2-16. The property damage was estimated at the equivalent of over 5 million U.S. dollars.

2.4.4.2 Kenya:

The hail climatology of the Kericho-Nandi area in Kenya has been reported recently by Henderson et al. (1970). This area may possess the highest incidence of hail in the world.

The persistent weather pattern consists of mostly clear mornings followed by small cumuli development over the high country before noon. By 1400 these clouds develop vigorously. As the westerly surface winds increase, the developed cumulus clouds move westward across the tea estates. Storm development over this area subsides after the late afternoon and the cycle repeats.

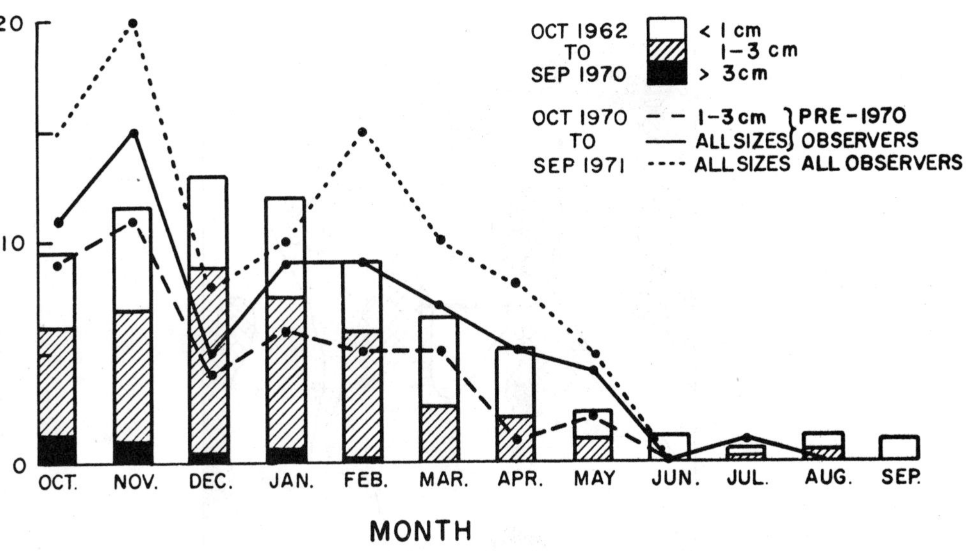

FIG. 2-15 Seasonal variation of hail days within the reporting network. Monthly averages for 1962 to 1970 and monthly totals for 1970/71 are given. Sizes refer to estimated diameters of the largest hailstones. (from Carte and Held, 1972)

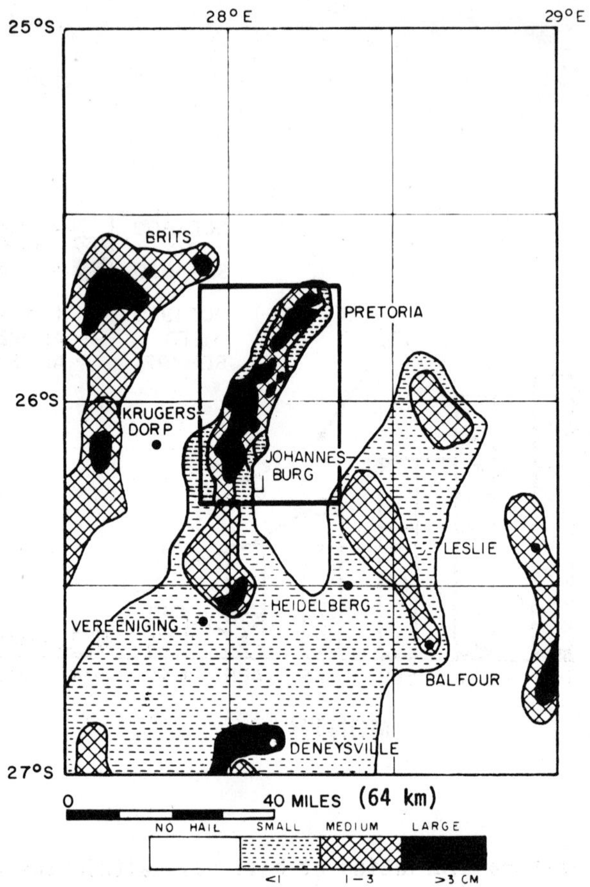

Fig. 2-16 Distribution of hailstone sizes; hailstorms in South Africa on January 15 and 16, 1964. (from Carte, 1964)

An average annual rainfall in the area of Kericho-Nandi Hills is about 177 cm. Hail reports indicate that more than 85 percent of the individual thunderstorms contain hail sometime during their life cycles. The number of hailfalls per estate during the 1967-1968 season is shown in Fig. 2-17. The great variation in number of hailfalls from place to place is apparent, the number varying anywhere between a single instance and 25. There is no evidence of a particular estate getting more hail year after year. Although the hail records are rather limited, the highest frequency of hail is during February and March, and again in August and September. However, hailfall is likely to occur during each month of the year somewhere in this area.

2.4.5 India, Pakistan and Bangladesh:

Remarkably detailed hail data are available for these areas from the crop damage (Ramdas et al., 1938). At least until recent years, the governments remitted land tax in part or in whole for the area affected. This practice made it necessary for a revenue officer to visit the affected area and make a detailed report to file with the collector of revenue. The data on crop damage show that almost all of the hailstorms in these regions occur during the northeast or dry monsoon; very few are recorded in the season of the southwest or wet monsoon. The first half of the year, January to June, includes most of the hailstorms in these regions with a maximum frequency in April. Very few

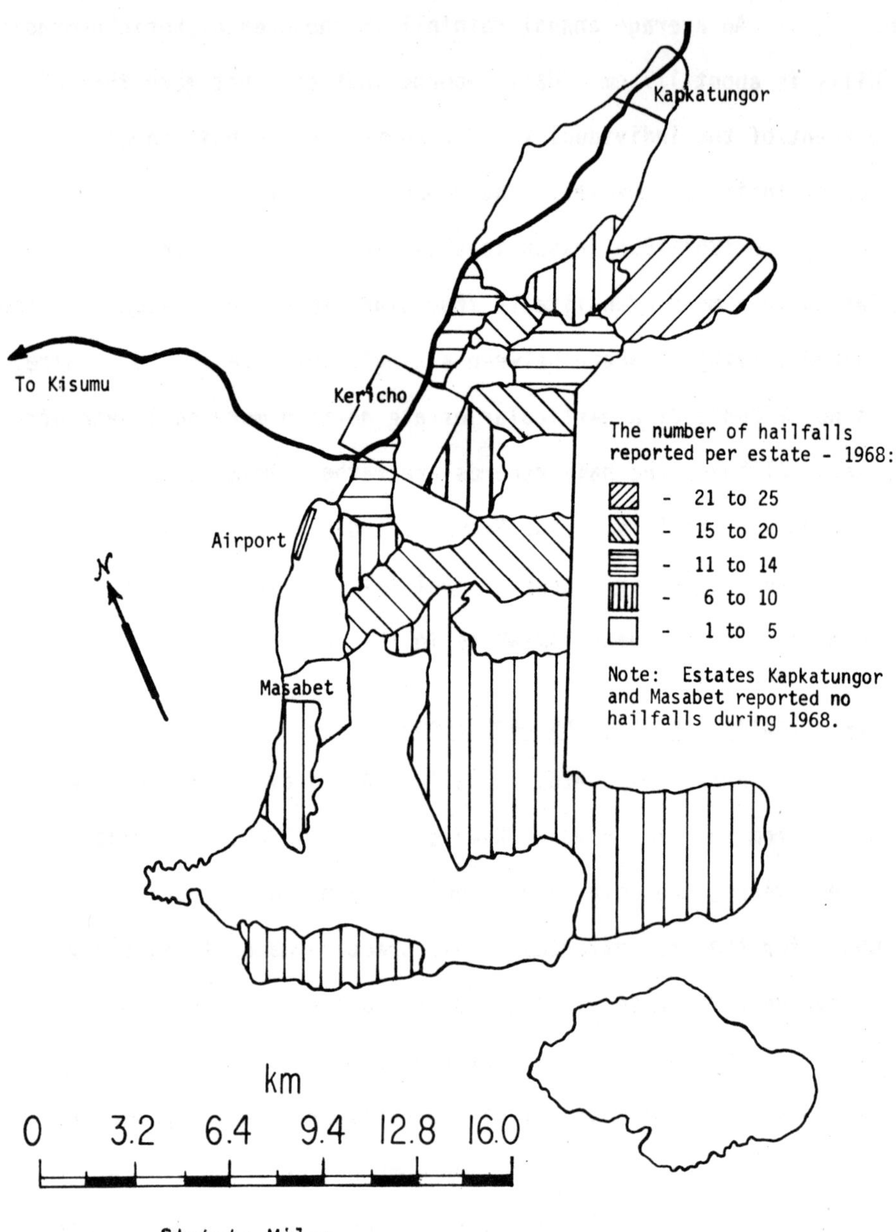

FIG. 2-17 Number of hailfalls per estate - 1968.

(from Henderson et al., 1970)

hailstorms occur from July to December (July to September is the monsoon period).

Ramdas et al. report that over the Western Himalayan region, the highest annual frequency is up to ten hailfalls per year (Fig. 2-18). The frequency decreases to five per year in the Eastern Himalayan region and the North West Frontier in Pakistan. One hailstorm per year is common to the areas of lower Bengal, Central Provinces, United Provinces and Bihar. This rate decreases to once in two years over the adjoining plains. Hailstorms are comparatively rare in the coastal regions. Relatively recently published data (Frisby and Sansom, 1967) for the region confirms the increase of frequency during the premonsoon period, culminating in a maximum frequency in the months of March and April.

2.5 Time of onset of hailfalls:

Hail frequency has been observed by many investigators to be greatest in the mid- or late afternoon, the period of maximum instability. A large amount of data has been reported by Flora (1956) on the time of onset of hailfalls from all over the USA. He found nearly three-fourths of the hailstorms occurred between 1600 and 1900 hrs and only a small percentage between midnight and midday (refer to Fig. 2-4).

Beckwith (1957) reports data on the time of hail onset as the percentage of hail reports for each hour. These data were

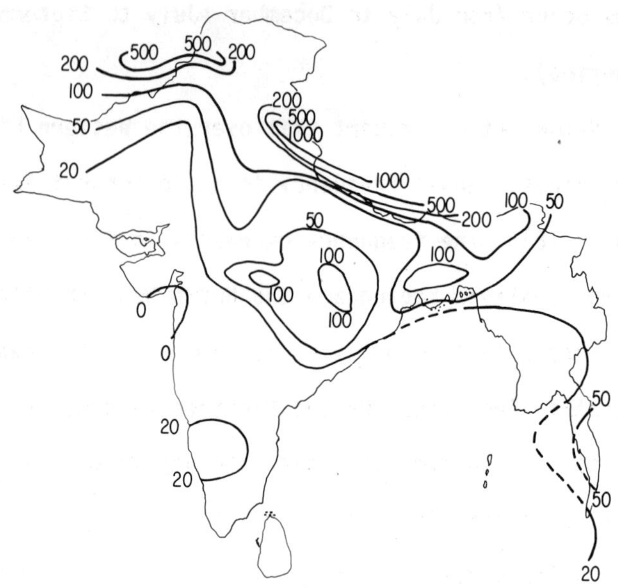

Fig. 2-18 Frequency of hailstorms in 100 years. (from Ramdas et al., 1938)

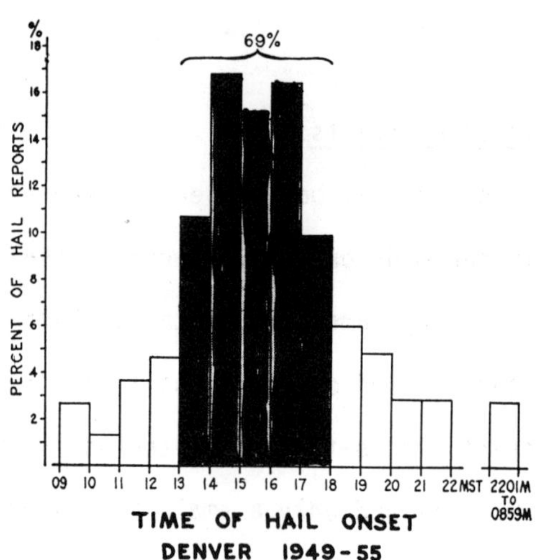

Fig. 2-19 Time of hail onset shown as percentage of total reports for each hour. This graph is based on 525 individual reports covering the 7-year period 1949-55. (from Beckwith, 1957)

collected in Denver and vicinity utilizing a network of 40 reporting stations. The histogram shown in Fig. 2-19 is based on 525 individual reports covering the seven year period 1949-1955. The optimum time for the hailstorms to begin is from 1400 to 1700 hrs, while 69 percent of the storms started between 1300 and 1800 hrs. Thus, the most frequent time for the hail occurrence reported by Beckwith is somewhat earlier than that reported by Flora (1956).

In Alberta, Canada, most of the hailstorms occur during the months May to September. A marked seasonal trend for time of hail onset in this region has been reported by Summers and Paul (1967). A more pronounced peak earlier in the afternoon was noted as the season progressed. Carte and Basson (1970), however, did not notice such a tendency for the hailstorms in Transvaal, South Africa. In Alberta, the fall of the largest hailstones increased toward the late afternoon, and the onset time also shifted toward the late afternoon as the distance from the Rocky Mountains increased.

Sulakvelidze (1967) has reported that in the Northern Caucasus, Poland and the central portion of European USSR, hail usually falls between 1300 and 1900 hrs. Hail was not observed to fall even once during 0200 and 0600 hrs. On the other hand, in Transcaucasia hail falls at any hour, and in eastern Georgia about 3 percent of the hailfall occurred between 2300 and 0700 hrs. In Armenia, hail falls between 2200 and 0200 hrs. Thus, in some regions there is no clearly pronounced relationship between the frequency of hailfall and particular hour of the day or night.

The distribution curves for the onset time of hail as reported by Carte and Basson (1970) are shown in Fig. 2-20. Sixty-six percent of the cases were reported between 1400 and 1900 hrs, which agrees reasonably well with the reports of other investigators. Carte and Basson point out that the data on hail occurrences at night may have some bias since the cases may not be reported as conscientiously as those during the day; this bias would be applicable particularly to less severe night storms. Figure 2-20(b) shows that the curve for the onset time of large hailstones is much flatter than the curve for smaller hailstones, indicating that severe storms are more likely to occur at any time of the day or night than mild ones. Seasonal differences were found to be small as apparent from Fig. 2-20(c).

2.6 Duration of hailfalls:

A wide range in the duration of hailfalls at a point was noted by Beckwith (1957) from the study of 450 hail reports from the Denver area and vicinity. The minimum duration for short bursts was 10 seconds, whereas there was one report of continuous hailfall over a period of 45 minutes. The most common values of hail duration were between 1 and 15 minutes, the median value being 5 minutes. No correlation was found between the size

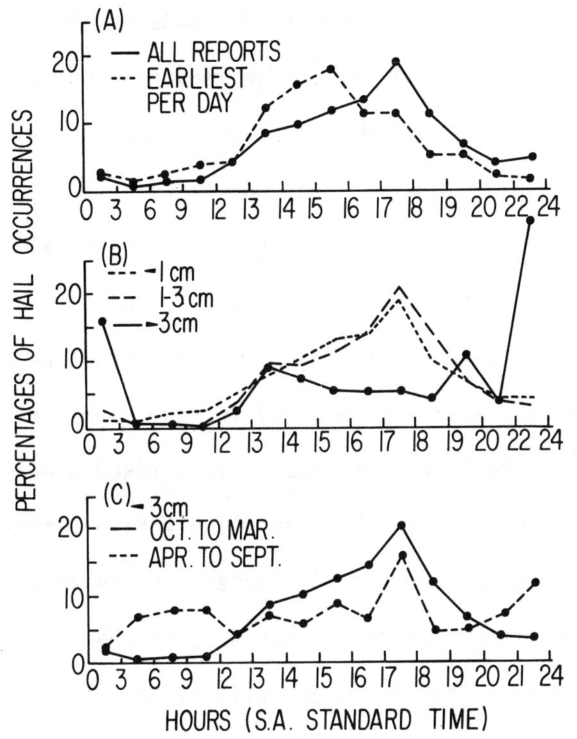

FIG. 2-20 Distribution curves for onset time of hail. (from Carte and Basson, 1970)

pattern and the duration of hailfall.

Table 2-9 is a summary of data on the regional variatio of the duration of hailfall. The median value varies between 2 and 15 minutes. Colorado and Alberta occupy very similar positions in relation to the Rocky Mountains. The storms originating in these areas and traveling over the continental interior have o the average, longer hail durations at a point.

Sulakvelidze (1967) has also summarized the data on duration of hailfalls (Table 2-10) in many countries. The time during which hail inflicts damage at a fixed place is taken as th duration of hailfall. The mean value varied between 5 and 10 minutes. The maximum was 90 minutes. No relationship was apparent between the duration and the size of the hailstones, confirming the earlier finding of Beckwith (1957).

The mean duration recorded at a point in more than 7800 reports in Transvaal, was 8 minutes (Carte and Basson, 1970). Th median value was 6 minutes. The duration was less than 3 minutes in nearly one third of the cases and longer than 12 minutes in only 17 percent. The longest duration was 45 minutes. Figure 2-21 shows the frequency distributions of the various durations. Continuous hailfalls and intermittent ones had durations of 7 and 10 minutes, respectively. The mean duration of continuous falls with bigger hailstones was 13 minutes, whereas the value is reduced to 5 minutes when only small hailstones resulted.

TABLE 2-9

Hailfall duration in various regions.

(from Paul, 1968)

Region	No of reports	Period covered	Frequency distribution	Median duration (minutes)
Bedfordshire, England	251	Unspecified	50% of falls 3 min or less; 2% over 30 minutes	2
New England	232	1956-58	Range from 0 to 25 min	3-4
Denver	450	1949-55	From 10 sec up to 45 min; falls up to 15 min common	5
Central Illinois	738	1958-62	No information	5-7 Mean 6-9
Northeastern Colorado	1,390	1960-64	63% less than 15 min; 10% longer than 25 min	10-15
Central Alberta	28,577	1957-66	73% shorter than 12 min; 7% longer than 25 min	7-12 Mean 10

TABLE 2-10

Duration (min) of hailfall at a point according to different authors.

(from Sulakvelidze, 1967)

Author	Location	Minimum	Maximum	Mean
Prohaska	Austria	0.6	50	8-10
Gigineishvili	Eastern Georgia	5-10	-	-
Beckwith	Mountain regions, USA	0.6	45	5
Defur	Belgium	5	20	-
Tverskoi	Rostov Region	5	20	10
Pastukh, Sokhrina	European part of the USSR	2	20-30	15
Chepovskaya	Northern Caucasus	3	30	5-10
Jénève	France	1	90	5-10
Weickmann	USA	-	85	5-10
Mean				6-7

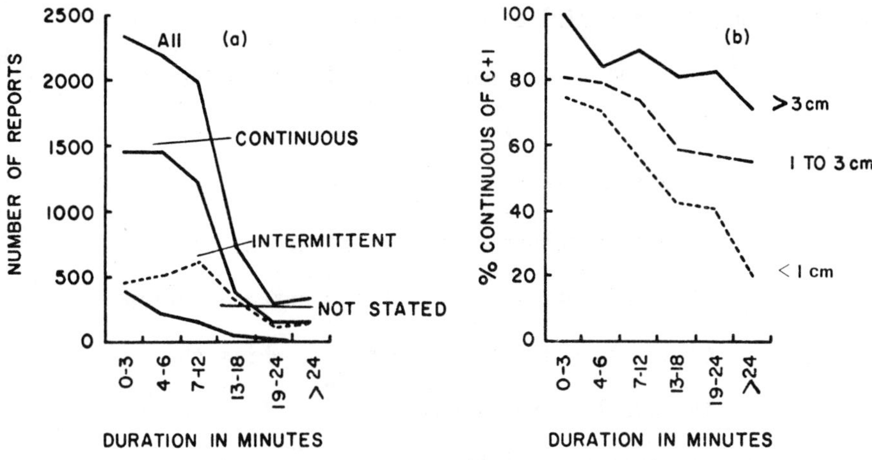

FIG. 2-21 (a) Number of reports in each of the various duration groups.

(b) Percentage of reports in each size group which stated that hailfall was continuous for various durations - where size refers to the largest hailstones.

(from Carte and Basson, 1970)

Details of intermittency of 272 falls as described by Carte and Basson (1970) are quoted below:

> "Up to six bursts of hail were observed. Three bursts were much less common than two - only 16 percent of these reports gave the number of bursts as three, 5 percent as four, and 2 percent as five. The average duration of all bursts was four minutes, while the average lull was eight minutes. There were big deviations from these mean values, the deviations being, to some extent, related to the total duration of the hailfall and to the size of the largest hailstones. The bursts became longer and the gaps shorter with an increase in the size of hailstone. Both gaps and bursts tended to be longer as the total duration increased."

2.7 Amount of hail:

Rather limited information is available on the amount of hail reaching the ground. This quantity can vary within wide limits. Obviously, the longer the hailfall continues at one point, the greater the amount.

Weickmann (1964) reports 45 cm depth in the USA over 135 km^2 area; this hailfall lasted for 85 minutes during a severe hailstorm at Seldon, Kansas. Sulakvelidze (1967) reported 25 cm depth over a similar size area in the USSR. The average depth of fallen hail, however, was 5-15 cm in the Northern Caucasus and 8-10 cm in eastern Georgia.

TABLE 2-11

Comparisons of hailfall durations

(from Carte and Basson, 1970)

Duration (min.)		<6	7-12	13-18	19-24	>25
Frequency (%) :	Alberta	47	26	13	7	7
	Transvaal	58	25	9	4	4

The results reported by Carte and Basson (1970), however, show considerably less depths of fallen hail. Eighty-five percent of all hailfalls reported did not produce sufficient hailstones to form a continuous carpet. In 50 to 60 percent of the cases, the spacing between the hailstones varied from 5 cm to zero. They observed only occasionally a layer more than few cms in thickness. The intermittent falls lasting less than 3 minutes usually produced only widely scattered stones.

2.8 Concentrations of hailstones:

Ludlam and Macklin (1959, 1960) have reported hail concentrations versus sizes ranging from 1.0 to 6.5 cm in diameter. These were estimated from the reports of the cooperative observers at various places during the passage of two severe storms in England. These and other concentrations reported by Douglas (1960) have been plotted by Atlas and Ludlam (1960), as shown in Fig. 2-22.

The curve $N_D = 40 \exp(-2.27D)$ for particles with diameters greater than 1 cm with N_D in $m^{-3}cm^{-1}$ and D in cm is in reasonable accord with the maximum concentrations which are found at the ground. Curves at the upper left in Fig. 2-23 are several size spectra of hailstones on the ground as reported by Douglas (1960). As given, they are spatial concentrations. They were no doubt derived from the number of hailstones per unit area, making use of the duration of hailfalls and the fallspeeds of the hailstones. No simple analytical law could be fitted to all the curves. However, all distributions exhibit similar features with respect to the range of diameters of hailstones, in spite of the hundred-fold variation in the precipitation rate. The small hailstones suffer significant losses due to melting during fall and further melting on the ground. When these losses are accounted for, it may be possible to extend the portions of the curves to small diameters to give exponential distribution.

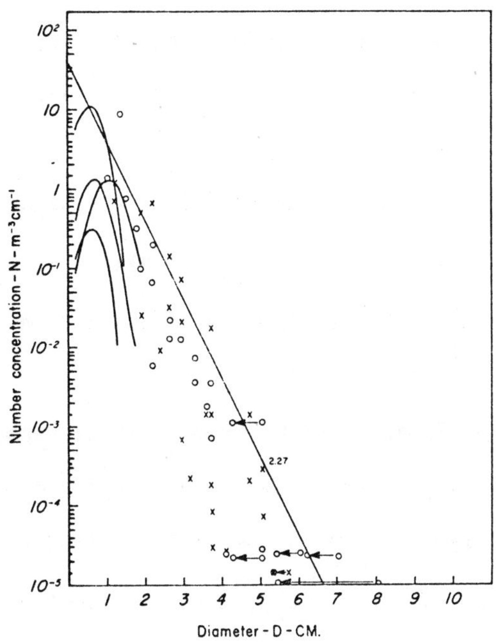

G. 2-22 Concentrations of the largest hailstones observed at various places on the ground in the Horsham [circles] and Wokingham [crosses] storms [in England] and an exponential curve fitted to the maximum values. For non-spherical stones, arrows lead to points representing the diameter of spherical stones of the same mass. Values plotted represent those deduced for the -10°C level, after assuming melting in a downdraft equal to the fallspeed of the stone. Curves at upper left are several size spectra of stones at the ground as reported by Douglas [1960]. (from Atlas and Ludlam, 1960)

The hail concentrations in New England have been reported by Donaldson et al. (1960). Apparently these show no consistency with size, but the median concentration does decrease from 0.14 to 0.07 m^{-3} in 0.5-1.0 and 1.5-2.0 cm diameter size groups, respectively.

The average graupel and hail size distribution observed in convective cloud systems over the High Plains in U.S.A. has been reported by Auer (1972). The mean distribution of diameters of hail is given by N_D=561.3D $^{3.4}$.

Recently, Federer and Waldvogel (1975) reported hailstone size distributions from a Swiss Multicell storm. The mean distribution for diameters of few mm to 25 mm is given by N_D=12 exp (-0.42D), which is in good agreement with spectra measured by Douglas (1963) and Ludlam and Macklin (1959).

2.9 Summary:

The climatology of hail in the USA is very diverse. In many states the active season is May, June, July and August, and the total loss per year due to hail damage is about $315 million. Hail damage in Alberta, Canada constitutes a serious problem for agriculture and the estimated direct losses of field crops are $23 million annually. The areas of major hailfalls in Western Europe are ascertained from the insurance data on yearly hail damage. The data in this region are, however, limited and it seems that more effort towards data acquisition would be worthwhile. The hailfall is also significant in Argentina, the Soviet Union, and certain parts of Africa. Kenya may possess the highest incidence of hail in the world and hailfall is likely to occur during each month of the year somewhere in this area.

The data on the time of onset of hailfalls from the countries all over the world show that nearly 75% of the hailstorms occur between 1300 and 1900 hrs. In contrast to this, the results published on duration of hailfalls are in a form that makes comparison difficult. However, frequency distributions of durations at different places are rather similar. All reports indicate a decrease in frequency with increase in duration. Detailed investigations have been conducted in Alberta and Transvaal, and the results are remarkably similar (Table 2-11). The limited information available on the amount of hail reaching the ground shows that it can fluctuate within wide limits. The concentrations of hailstones vary between 1 m^{-3} to 10^{-4} m^{-3} for diameters 1 cm to 5 cm.

CHAPTER 3

FALLOUT PATTERNS OF HAILSTORMS

It has been well recognized that a synoptic network of weather observing stations in any country is too scattered to give any realistic hailfall patterns in a given region. Reports of hail from a dense network of observers on the ground provide invaluable detail about its surface pattern. Detailed hail and rain data from such networks have been used by a number of groups to define and study hailswaths and hailstreaks and their associated rainfall. Four such well-organized networks, the first in the Denver area, the second in Central Illinois, the third in Alberta, Canada, and the fourth in South Africa, will now be described. Two other networks, not so well organized but from which have come reports on some characteristics of hailfall, were located in NE Colorado (Schleusener and Auer, 1964) and in New England (Donaldson et al., 1960).

3.1 Hail reporting networks:

3.1.1 Denver area:

The characteristics and climatology of hailstorms in the Denv area have been reported by Beckwith (1957, 1960). A network of 40 reporting stations (Fig. 3-1) was utilized to collect the data over a period of 10 years (1949-1958). Black circles are the reporting points

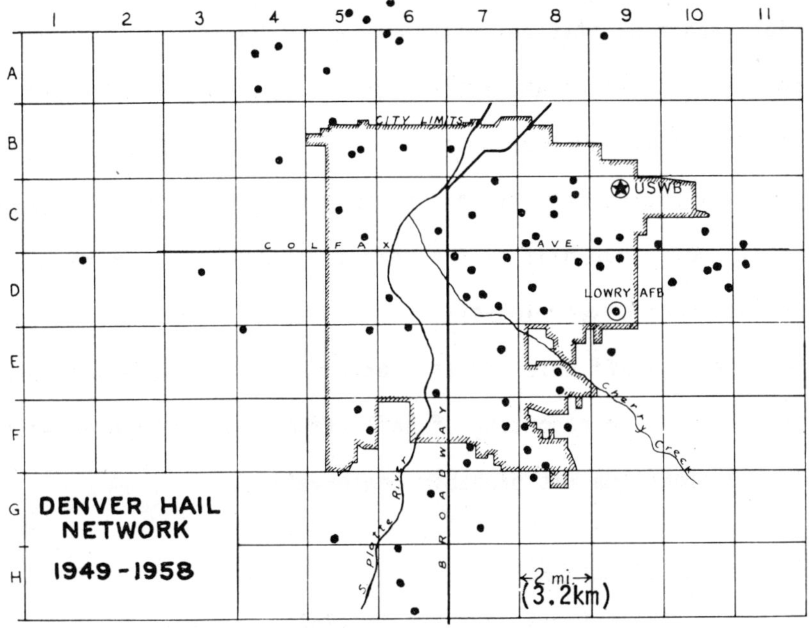

FIG. 3-1 Hail reporting network in Denver area; the Denver rawinsonde release point is just east of U.S. Weather Bureau [now National Weather Service] station. (from Beckwith, 1960)

which existed for one or more hail seasons. There were unofficial reporting stations within or near the city of Denver. A square grid system was superimposed on the region with each grid sector containing four square miles. This was assumed to be an ideal reporting density for determining areal distribution of hail. More than half of the reporting stations were manned by professional meteorologists. In addition, nearly one-third of the hail study staff were either Weather Bureau employees or airline personnel. The reporting of hail from storms which developed during nighttime was probably conservative.

3.1.2 Central Illinois:

The network operated in 1967 in Central Illinois included 49 recording raingage-hailpad sites plus 86 cooperative observers extending over an area of 400 mi^2. Analysis of the data from this network indicated the necessity for a larger and denser network, which was established by the Illinois State Water Survey in 1968 (Fig. 3-2). It contained 196 recording raingauges, 100 hailpads and 261 cooperative hail observers. The Dense Network, which was located within the above mentioned total network, had 96 hailpad sites in a 100 mi^2 area. In both of these networks, the density of the fixed hail observation points was 1 site per 3 mi^2.

The hailpads were made of styrofoam one inch thick and one square foot in area. They were wrapped in aluminum foil and mounted on a supporting platform. Hail size was correlated with

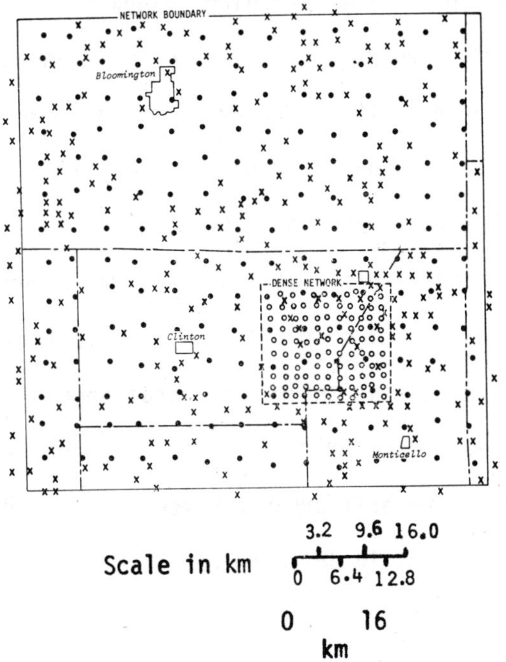

- ● Recording raingauge and hailpad station
- o Hailpad station
- x Cooperative observer

Fig. 3-2 Central Illinois hail-rain network operated in 1968. (from Changnon, 1970)

foil indentations. The recording raingages in both hail-rain networks were modified to record time of hail onset to the nearest minute (Changnon, 1966). Moreover, the data were supplemented with insurance loss data and on-the-spot field survey checks. The insurance companies furnished the details of the date of loss, location of loss within individual fields, amount of loss at each point and the type of crop affected. The data on time of hailfall, hailstone sizes, their concentration per unit area and the type of stones were obtained from the cooperative observers. The confirmation of no-hail areas was equally as valuable as the knowledge of hail areas in defining hailstreaks.

3.1.3 Alberta, Canada:

The Alberta Hail Studies (ALHAS) project is located in central Alberta, and was started in 1956. It consists of an area 63,000 km^2 which is bounded on the west by the foothills of the Rocky Mountains (Fig. 3-3), and is notorious hail country. East of the foothills lie slightly rolling farmland but the area is chiefly flat prairie, with mixed farming and cereal crops predominating.

An outline of the ALHAS field observational program during 1967 and 1968 is described by Chisholm and English (1973).

(a) Hailfall observations:

The data were collected with the cooperation of the farmers when hail fell on their land, either by a hail-report post card or by telephone surveys.

(b) Precipitation network:

A network of 540 stations over an area of about 31,000 km^2 was set as shown in Fig. 3-4. The average station density was approximately 1 per 57 km^2. A "high density" area of 1200 km^2 contained an additional 97 stations with an average density of 12 km^2. In the Rocky Mountain Forest Reserve, sparse population and poor roads limited the station density near the western boundary (Deibert and Renick, 1975).

(c) Weather observations:

The weather stations located at four places (see Fig. 3-3) supplied hourly weather reports by teletype, and radiosonde soundings were taken twice daily at Edmonton and at Penhold.

(d) Radar observations:

A modified AN/FPS-502 radar operating at wavelength 10.4 cm was used to obtain a five-level gray-shade PPI display which was photographed with 35-mm film for later study.

(e) Aircraft observations:

An instrumented C-45H aircraft was operated to obtain cloud base observations of vertical velocity, updraft area, temperature and ice nuclei concentrations.

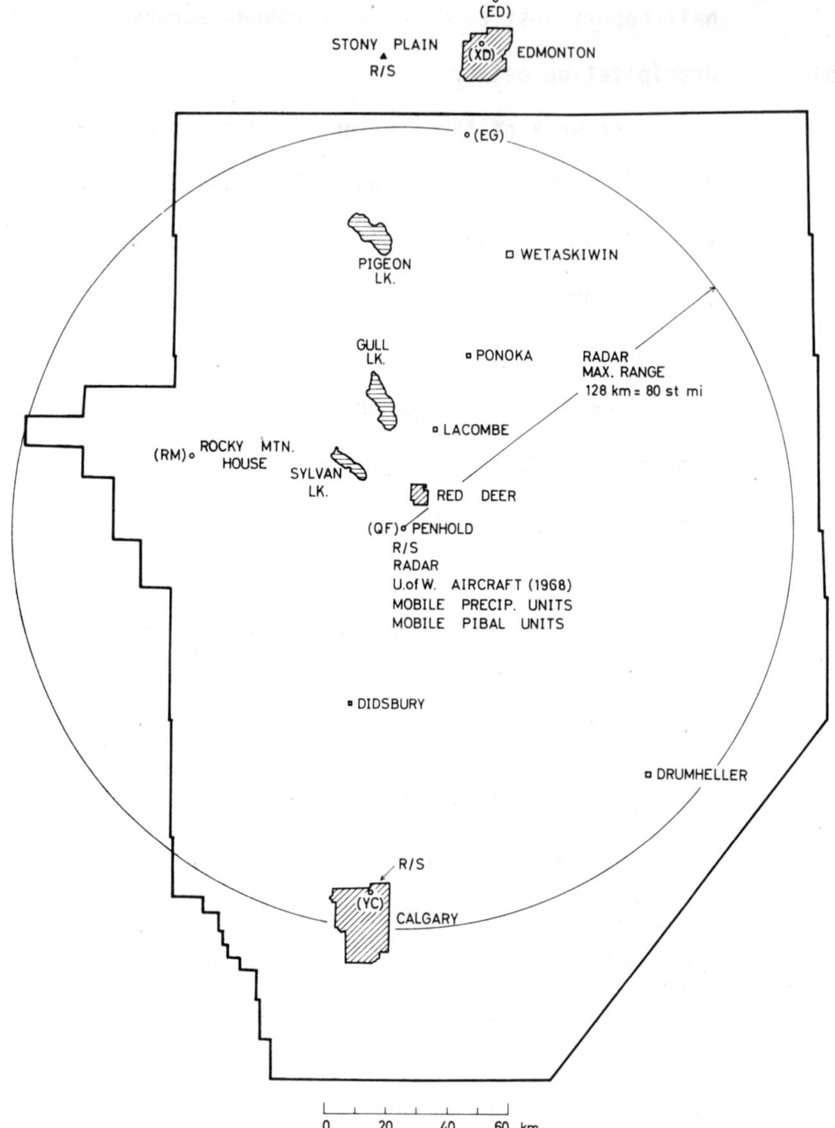

FIG. 3-3 Map of Alberta Hail Studies (ALHAS) project area. The project area boundary is indicated by the heavy solid line. Atmospheric Environment Service weather reporting stations are shown by small circles and radiosonde stations are marked (R/S). Observing equipment located or based at project headquarters is also indicated. (from Chisholm and English, 1973)

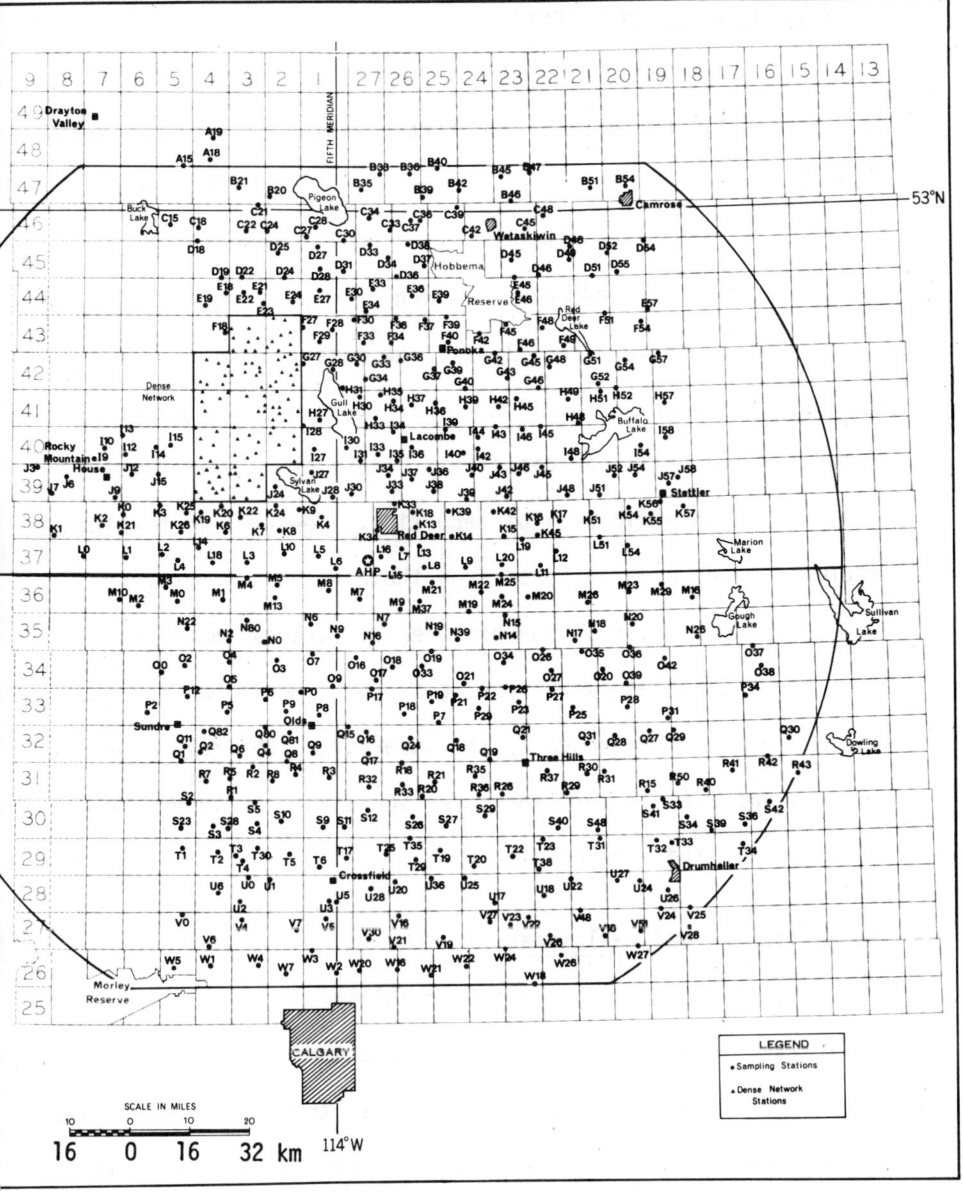

LEGEND: • Sampling Stations ▲ Dense Network Stations

FIG. 3-4 Precipitation sampling network in Alberta, Canada. (from Deibert and Renick, 1975)

3.1.4 Transvaal, South Africa:

Carte and Basson (1970) report the details of hail climatology of the Pretoria-Witwatersrand region of the Transvaal, South Africa. The data were gathered with the help of voluntary observers. A smaller network over a 128 km^2 area was operated for 10 years. The expanded network which covers an area of 2560 km^2 has been in operation since 1962. The number of voluntary observers varied between 700 and 1000, which were neither uniformly distributed nor equally conscientious. The storm severity was a major factor in efficiency of reporting. However, much valuable information has been gathered from these voluntary observers.

A facsimile of the hail report card is shown in Fig. 3-5. Eight thousand such cards relating to more than five hundred days of hail were returned by the observers. The reporting network was improved in 1970 when many additional observers were enrolled within the same area.

3.2 Hailswaths and hailstreaks:

An outstanding network established by Prohaska (1900, 1907) extended over the whole of Austria and recorded hail damages since 1885 to the beginning of the twentieth century. His published results are of unique value to everyone interested in the studies of hail. He found that frontal hailstorms were more severe than the air mass type. The hail paths were hundreds of kilometers long and 10-20 km wide. Fig. 3-6 shows several such

GAcsir 114d 74

HAIL REPORT

MARK APPROPRIATE SQUARES

NAME: .... PHONE: ....

HOME ADDRESS: ....

LOCATION OF HAIL: ....

DATE OF STORM: .... HAIL BEGAN: .... a.m. p.m. HAIL LASTED: .... minutes

Your time of hail beginning is correct within 1 min. ☐ 2 min. ☐ 5 min. ☐ 10 or more minutes ☐

SIZE OF HAILSTONES:	Rice	Pea ¼"	Grape ½"	Walnut 1"	Golf ball 1½"	Hens egg 2"	Tennis ball 3"	IF LARGER, GIVE DIAMETER
LARGEST	☐	☐	☐	☐	☐	☐	☐	....
SMALLEST	☐	☐	☐	☐	☐	☐	☐	....
MOST COMMON	☐	☐	☐	☐	☐	☐	☐	....

Hailfall was : continuous ☐ intermittent ☐ Hailstones were hard ☐ soft ☐

Hail started : before ☐ after ☐ at same time as ☐ rain

Largest hail started : before ☐ after ☐ at same time as ☐ smaller hail

Average distance between hailstones on ground .... inches OR depth of hail on ground .... inches.

REMARKS: ....

....

Hail preserved? Yes ☐ No ☐ More cards required? Yes ☐ No ☐

FIG. 3-5 Hail report card supplied to observers.
(from Carte and Basson, 1970)

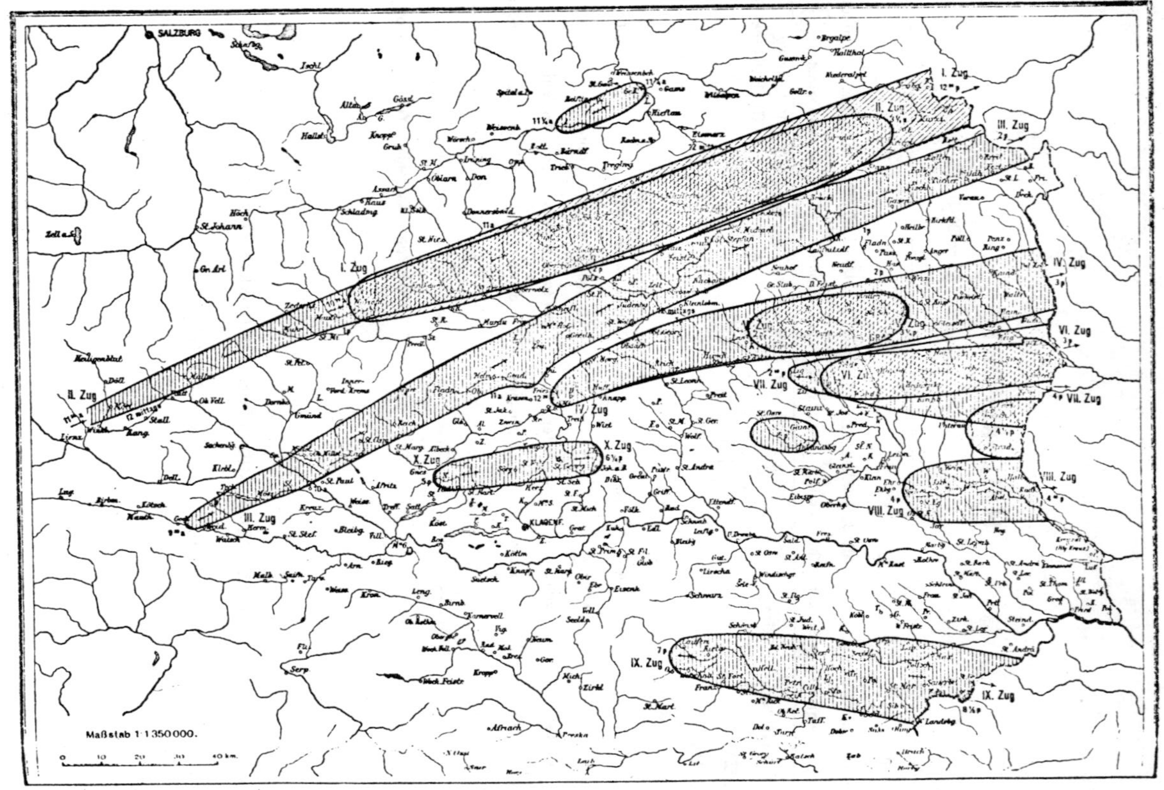

FIG. 3-6 Hail paths of July 6, 1905, in Austria, according to K. Prohaska.

(from Weickmann, 1953)

hail paths. Preferred routes did exist and their widths changed in the direction of movement. The broadening of the hail path was normally accompanied by heavier hailfall. In the central part the hail was more severe. However, often small areas within such a hail path were found to be completely devoid of hail.

Detailed distribution of maximum hail size in a severe storm in S.E. England reported by Ludlam and Macklin (1959) is shown in Fig. 3-7. Though the data in many cases were obtained on several succeeding days by asking inhabitants about the details of the storm, it was possible to construct a coherent set of isopleths of the maximum sizes of hailstones. The fallout was intermittent and not steady. The isopleth patterns in Fig. 3-7 consist of strips lying along the wind direction. The big hailstones fell over large areas due to overlapping of such strips. The typical period of heavier hailfalls was about 5 minutes. It was not possible to deduce accurately the growth period of the large hailstones from the isopleth pattern and wind speed.

The data collected in Alberta by Douglas and Hitschfeld (1959) on hailswaths and radar echo envelopes showed a close relationship. Hailswaths were quite variable in total length and hailfalls in total duration; these values were 50 miles as median length and 2 hours as median duration (Fig. 3-8). The longest swath was in excess of 120 miles and lasted 4 1/2 hours. Though there was gross continuity to the hailswaths, there were regions

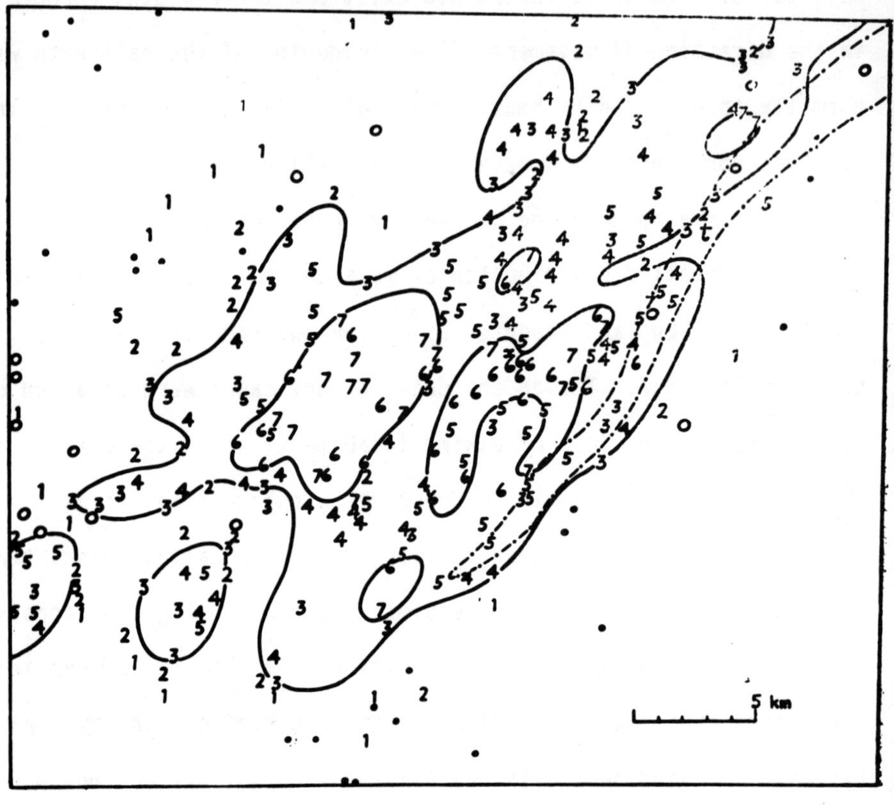

FIG. 3- 7 Detailed distribution of maximum hail size in the Horsham area. Here the numbers represent hail size according to the following scale:

0 - no hail; (a small dot in the diagram indicat some hail of unspecified, probably small siz

1 - pea (diameter about 0.6 cm)

2 - marble (1.6 cm)

3 - walnut (2.5 cm)

4 - golf-ball (3.5 cm)

5 - hen's egg (4.5 x 5.5 cm)

Continued next page

FIG. 3- 7 Continued

6 - tennis ball (7 cm)

7 - half grapefruit (9 cm)

The dots show where it is known that some rain fell without hail. The isopleths within which stones of diameter greater than 2.5 cm (size 3) or 7 cm (size 6) fell are drawn; the pecked line on the right shows the approximate area within which tree damage by tornado winds occurred. At the point marked 't' a brick-works in an open field was destroyed. (from Ludlam and Macklin, 1959)

from which no positive indication of hailfall was received.

The use of automobiles to track radar echoes with radar reflectivity greater than 10^5 mm^{6}m^{-3} has been found by Browning et al. (1968) to provide a workable technique for collecting large hailstones. The persons interviewed in the farming communities in Oklahoma gave detailed and consistent accounts of the storms. The resulting distribution of maximum hail size is shown in Fig. 3-9. It was centrally located and the hail size decreased on both flanks. The areas with radar echoes for which the radar reflectivity was greater than 10^5 mm^{6}m^{-3} were fairly closely related to both the swath of hail and heavy rain.

The swaths studied by Carte et al. (1963) were more or less straight paths with edges rather ill-defined. The largest hail was located at the center of the swath; the hailstone sizes diminished regularly to either side of the center line. There was sufficient evidence to indicate that the patchiness of the hail pattern is characteristic of all the hailswaths examined. This is in perfect agreement with data on crop damage (Changnon, 1970) and observations taken with the airborne infrared radiometry (Goyer and Roads, 1971).

Carte (1964) reports the hailfall from storms in Johannesburg, Pretoria and surroundings on January 15 and 16, 1964. The area covered was 350 square miles and there were in all 244 hail reports. With these data, it was possible to demarcate

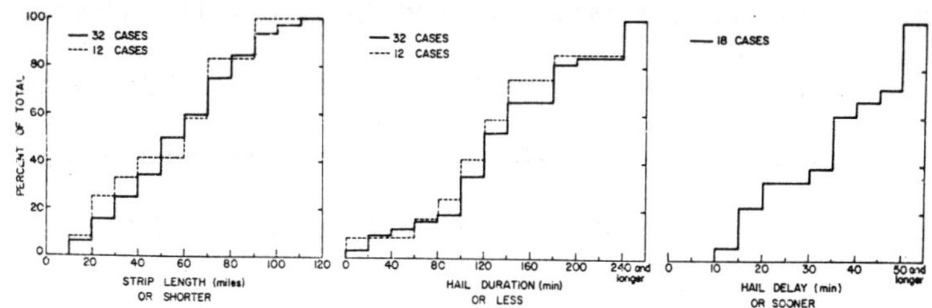

FIG. 3- 8 Cumulative histograms (a) of the lengths of hail strips, (b) of their duration. Here 12 of the 32 strips were particularly well defined and continuous. (c) Time delays between first radar echo and first hail observation. (from Carte et al., 1961)

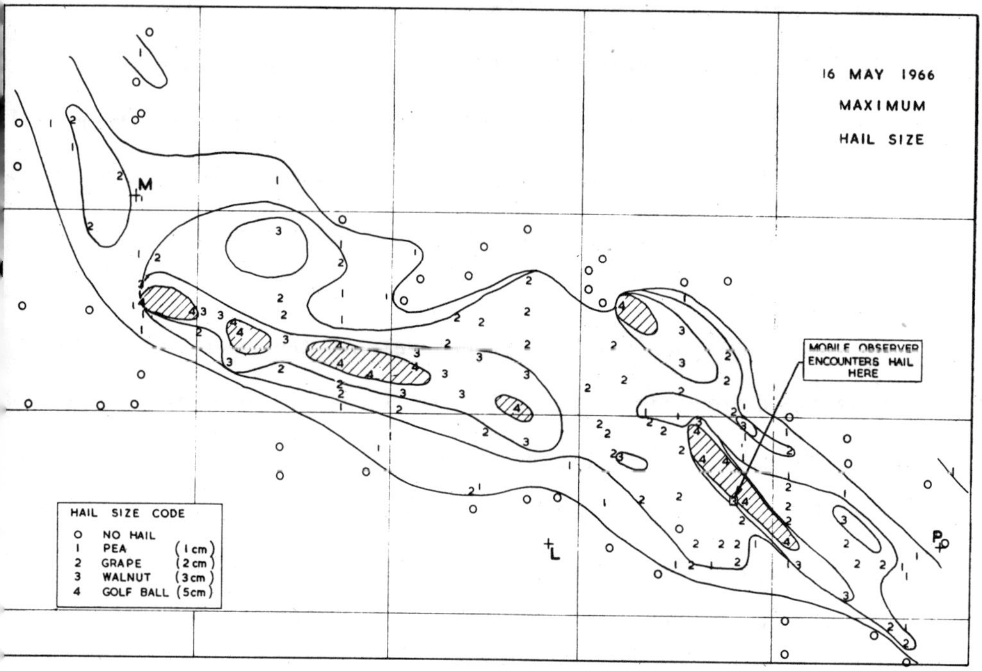

16 MAY 1966
MAXIMUM
HAIL SIZE

MOBILE OBSERVER ENCOUNTERS HAIL HERE

HAIL SIZE CODE
O NO HAIL
1 PEA (1 cm)
2 GRAPE (2 cm)
3 WALNUT (3 cm)
4 GOLF BALL (5cm)

FIG. 3-9 Maximum hail size, 16 May 1966. Individual observations are plotted according to the indicated code and areas receiving golfball size hail are hatched. (from Browning, Hallett, Harrold and Johnson, 1968)

the boundaries of the hail path and the isopleths of various hailstone sizes. Fig. 3-10 shows the hail path, divided into four strips, A B C and D. Largest hailstones were reported within the strip B and smaller hailstones became frequent in the strips A, C and D.

The Central Illinois dense networks operated during 1967 and 1968 have been described in an earlier section (3.1.2). Detailed hail and rain data have been collected with the help of these networks and analyzed by Changnon (1970). The major aim of the study was to determine the characteristics of hailstreaks and their associated rainfall. Changnon et al. (1967) have defined 'hailstreaks' as areas of hail continuous in space with temporal coherence. Each hailstreak most likely results from a single hail cell within a hailstorm. Therefore, detailed information on hailstreaks may be helpful in understanding storm dynamics and cell development.

Changnon's (1970) paper presents the results of extensive field investigations on 434 hailstreaks made during a two-year period. Hailstreaks resulting from a given hailstorm showed considerable variability in both frequency and physical characteristics such as size, duration, orientation motion, energy, etc. These characteristics are summarized in Table 3-1. Detailed maps showing the time, size and frequency of hailstones at all points are presented in Fig. 3-11. These plots also include no-hail

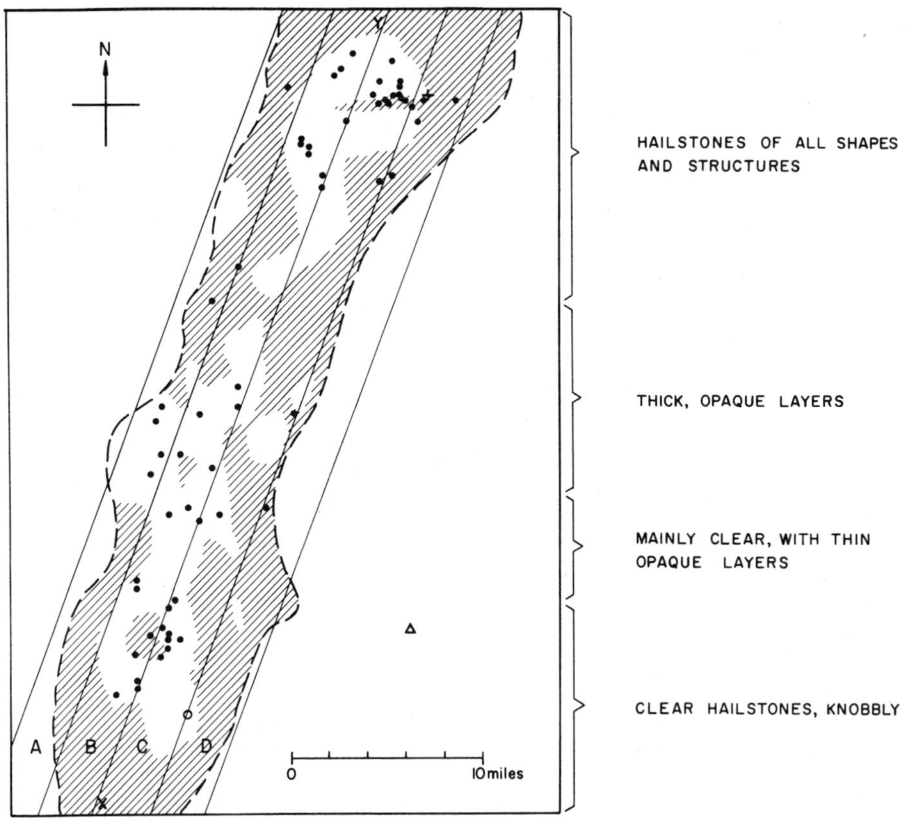

FIG. 3-10 Hailstone samples were received from 62 places shown by dots. Unshaded areas between the broken lines are where large hailstones (> 3 cm) fell.

o City Hall, Johannesburg

Δ Jan Smuts Airport

\+ Weather Bureau, Pretoria

(from Carte, 1964)

TABLE 3-1

Characteristics of hailstreaks in Central Illinois.

CHARACTERISTIC	Minimum	Maximum	Median	Remarks
Number per system	1	63	5	
Minimum distances between hailstreaks	24 km	38 km		
Area	2.3 km^2	2017 km^2	20 km^2	The correlation between area and mean energy was quite poor.
Mean energy ($joules/m^2$)	1.46×10^{-3}	185	8.9×10^{-2}	
Duration			10 min	
Speed			48 km hr^{1}	
Length			9.5 km	
Preferred orientation			264°	
Point frequency in a given hail season	0	10		
Point duration of hailfalls			3 min	Daylight hours - 4 min Nocturnal hailfalls - 2 min
Late afternoon hailfalls			43%	
Nocturnal hailfalls			28%	
Number of hailstones per hailfall per ft^2 (1 $ft^2 \simeq 0.09$ m^2)	6		24	Maximum number from a single hailstreak 1402

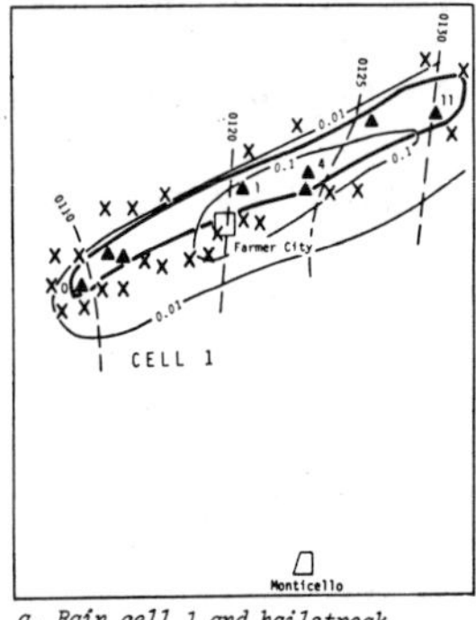

a. Rain cell 1 and hailstreak, 0110-0130 CDT.

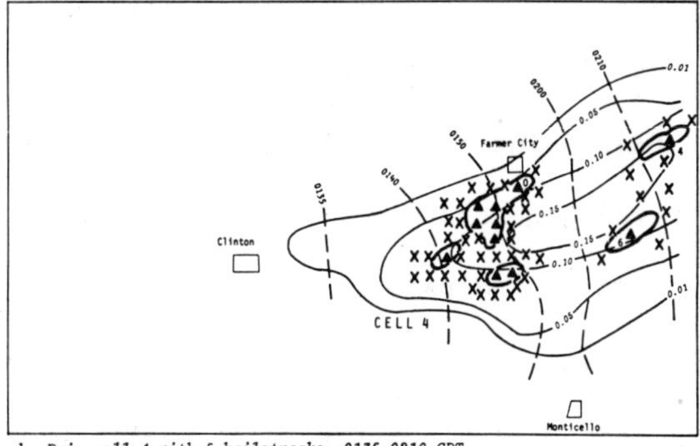

b. Rain cell 4 with 5 hailstreaks, 0135-0210 CDT.

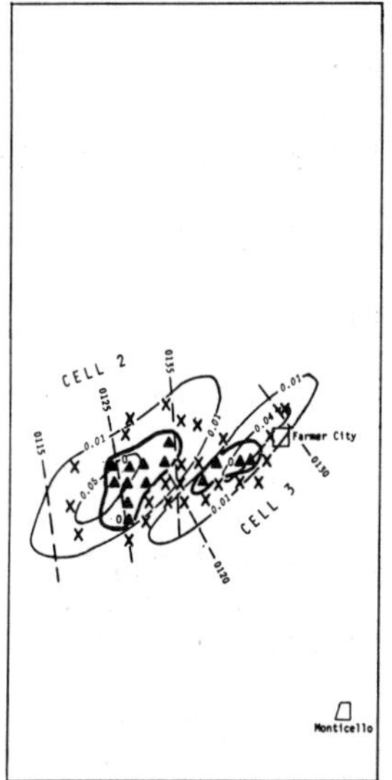

c. Rain cells 2 and 3 with 2 hailstreaks, 0115-0135 CDT.

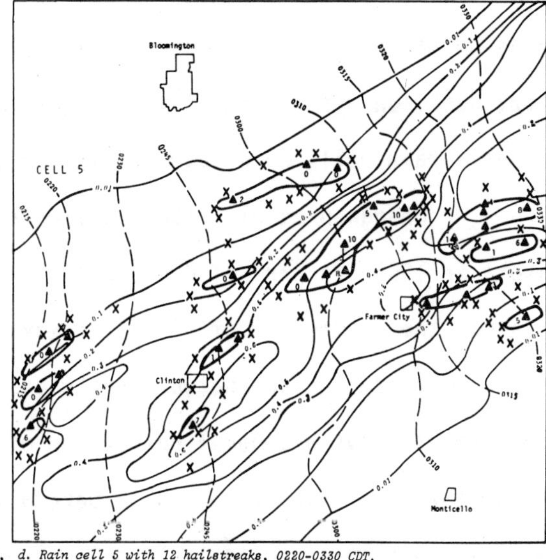

d. Rain cell 5 with 12 hailstreaks, 0220-0330 CDT.

EXPLANATION

X = Report of no hail with each rain cell

▲5 = Hail report, and where known, the time difference (minutes) between start of rain and start of hail

0.01 = Isohyet for total rainfall (inches) from rain cell

Hailstreak boundary

– – – Isochrones of rain start time, CDT

FIG. 3-11 Hailstreaks and rain cells on 1 July 1968. (from Changnon, 1970)

reports which were extremely useful in demarcation of the exact areas of hailstreaks. The amount and time (to the nearest minute) of rain associated with each hailfall were obtained by analyzing the raingage charts. These values were plotted as an isohyetal pattern to indicate the "rain cell." Figure 3-11 shows a 4100 km^2 hail-rain network area within which 20 hailstreaks occurred during a 2.5-hr period. These hailstreaks and rain cells reveal a wide variety of sizes and shapes; such patterns were typical of those found on many hail days. Occasional overlap of hailstreaks of short duration was noticed. The larger hailstreaks which occurred in heavy rain cells were responsible for most of the crop damages.

Changnon et al. (1967) collected field data in South Dakota in 1966, as part of Project Hailswath (Schleusener, 1966) to study the occurrence of 82 hailstreaks on 6 hail days. The lack of a dense surface network restricted somewhat the number of measurements, but a general comparison of regional differences in hailstreaks in Illinois and South Dakota was still possible. The results are summarized in Table 3-2. From these limited data it seems that the average hailstreaks in South Dakota are definitely wider in size and longer in duration than those in Illinois. Besides, a typical hailstorm in South Dakota produces fifty percent more hailstreaks. However, their speed, direction of motion and location with respect to the rain cell seem to be similar to those in Illinois.

TABLE 3-2

Comparison of results from Illinois and South Dakota.

HAILSTREAK CHARACTERISTICS	in SOUTH DAKOTA	in ILLINOIS
Average maximum length	24.5 km	9.4 km
Average maximum width	3.7 km	1.7 km
Maximum width	12.0 km	30.4 km
Greatest length	76.8 km	96.0 km
Speed (median value)	52.8 km hr^{-1}	48.0 km hr^{-1}
Preferred direction of motion	from the WSW (secondary maximum from the WNW)	from the WSW
Average number per hail system	14	9
Average duration	27 min	10 min

3.3 Intermittency of hailfalls:

McBride (1964) demonstrated with the improved surveys of Alberta hail patterns that the body of a hailswath does contain hail-free areas. Therefore the hail cells must have waxing and waning periods. Thus most of these storms possess a remarkable degree of intermittency which would be more consistent with a cellular storm model than a truly steady-state one. Similar results have been reported by Carte (1964). The distribution of hail indicated that several convective columns co-existed and produced hail in patches. Carte calls them 'cells.' In drawing the cells, reports (though insufficient in number) of intermittent hail were taken into account. The speeds of these cells were about half the speed of the cell-complex. The size and motion of these cells indicated a multi-cellular formation of the storm. Carte therefore concludes that there could not have been a continuous, unchanging updraft uniform across the whole area. There was the waxing and waning effect which better fits the changing outline of the hail cells and the patchiness of the hailfall.

Pell (1969) examined the changes in the character of the hailfall along a swath as time progresses--such as the maximum hail size, intermittency, soft hail size, point-hailfall duration and rain/hail sequence. His statistically significant results are shown in Table 3-3. Only a relatively small fraction of the total number of storms studied showed significant variations.

TABLE 3-3

Changes in the character of the hailfall, along the swath, or with time (Alberta, Canada). (from Pell, 1969)

Parameter	Variation	No. of storms examined	No. of storms with significant changes
1. Maximum hail size		47	
	Large to small		11
	Small to large to small		7
	Small to large		4
			22 (47%)*
2. Intermittency (no. of bursts)		29	
	Intermittent to continous		5
	Continuous to intermittent		1
	Continuous to intermittent to continuous		1
			7 (24%)
3. Maximum soft hail size		17	
	Large to small		3
	Small to large		2
			5 (29%)
4. Point-hailfall duration		42	
	Long to short		12
	Short to long to short		3
			15 (36%)
5. Rain/hail sequence		38	
	"Rain first" predominates		13
	"Hail first" to "Rain first"		2
	"Simultaneous" to "Rain first"		1
			16 (42%)
6. Time of fall, largest hail		32	9 (28%)

* In brackets, the percentage of storms examined in which the relevant parameter varied significantly.

Carte and Held (1972) compared the averages for intermittency of hailfalls in Transvaal and Alberta (Fig. 3-12). Though this comparison is certainly valid, limitations of the data should be pointed out. For example, a break in the hailfall may be a few seconds for certain reporters whereas others might have only reported breaks which were more than a few minutes long. One or two bursts per hailfall is more common. The frequency of having more than three bursts is rather low. From the detailed analysis, Carte and Held suggest different reasons for the intermittency of point hailfalls. These are movement of hail cells of irregular shape, repeated expansion and contraction of hail cells, and the passage of more than one cell over a given point.

3.4 Hailstreak-rainfall relations:

A typical example of the hailstreak-rain cell analysis reported by Changnon (1970) is shown in Fig. 3-11. Such a mesoscale analysis furnishes invaluable information on hail-rain relations including the occurrence of hail with respect to the core of a rain cell and the stage of cell development.

The following characteristics of hailstreak-rainfall relations are summarized from Changnon's (1970) paper.

(i) "Average dimensions for rain cells within the 1600 mi^2 network were 16 mi by 4 mi."

(ii) "Sixty percent of the hailstreaks

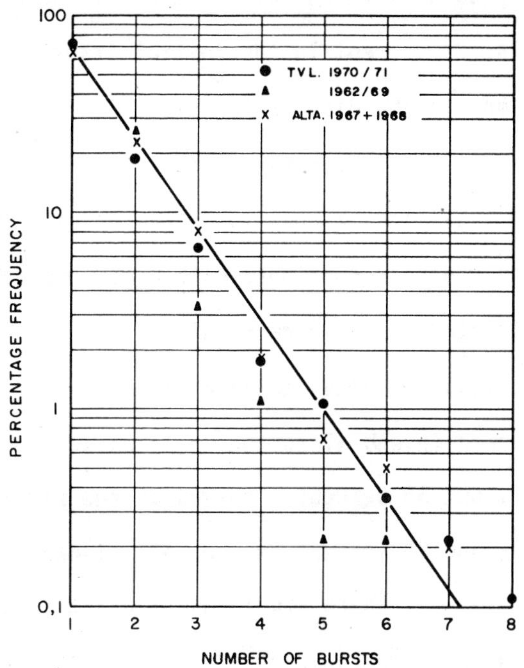

FIG. 3-12 Percentage report frequency against number of bursts for point hailfalls in the Transvaal and Alberta. (from Carte and Held, 1972)

occurred during the mature stage of their rain cells, 24 percent in the initial stage and 16 percent in the dissipation stage."

(iii) "In general, a rain cell produced only 1 hailstreak, but 20 percent of the rain cells produced 4 or more with a maximum of 25 associated with one rain cell."

(iv) "Forty-one percent of the hailstreaks occurred in and along the major axis of a rain cell, but 36 percent occurred on the right flank and 25 percent on the left flank."

(v) "There are frequent differences between the forward speed of the rain cell and that of its hailstreak."

(vi) "The median value of the average rainfall within a hailstreak was 0.19 inch; the maximum point value was 0.25 inch. The hailstreaks normally do not occur in the heaviest rain."

(vii) "The median rainfall rate in the hailstreak was 0.63 inch hr^{-1}."

A number of observers have reported on the sequence in which hail and rain first reached the ground. The results are summarized in Table 3-4.

TABLE 3-4

Observations of sequence of hail and rain.

OBSERVERS	REGION	SEQUENCE OF HAIL AND RAIN
Donaldson et al. (1960)	New England U.S.A.	Usually the hailfall started four minutes after the rain.
Williams and Douglas (1963)	Alberta, Canada	Every storm included all three categories; rain first, hail first, rain and hail together. When hail preceded rain, largest hailstones usually appeared early in the hailfall.
Sulakvelidze (1967)	Soviet Union	The sequence was rain followed by small hail and then large hail, when the wind was strong. It is not clear why the sequence reverses when the wind was light.
Carte and Basson (1970)	South Africa	Usually rain preceded hail; when hail arrived before rain, the hailstones were large; rain and hail commenced simultaneously in about one third of the cases. *
Changnon (1970)	Central Illinois U.S.A.	There were all three categories noted; rain first, hail first or rain and hail together. Sixty percent of the hailstreaks occurred during mature stage of rain cells.

* Figure 3-13 includes the histograms showing the sequence in which hail and rain reached the ground.

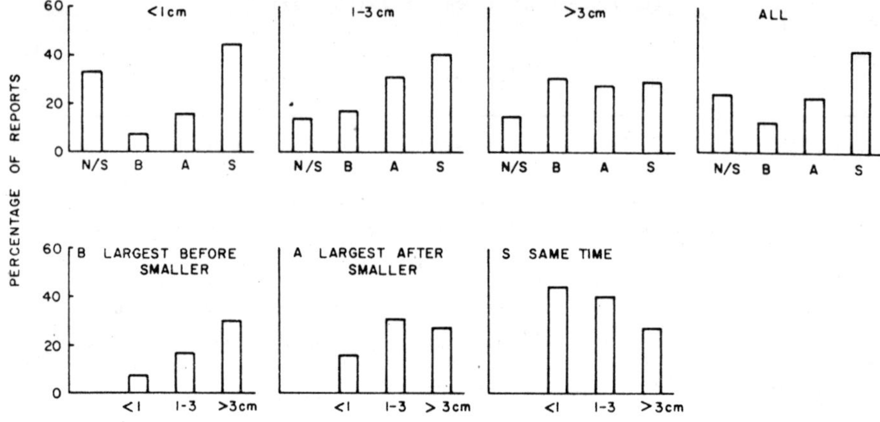

Legend

N/S = Sequence not stated

B = Hail before rain

A = Hail after rain

S = Hail and rain started at same time

FIG. 3-13 Histograms showing the sequence in which rain and hail reached the ground. [All percentages are of the total number of reports in a particular size group.] (from Carte and Basson, 1970)

3.5 Summary:

Reports of hail must come from a well-organized and dense network of observers to provide detail about its surface pattern sufficiently to define hailswaths, hailstreaks and the associated rainfall. In addition to hail observers, hailpads and recording rain-gauges are used.

The large hail paths may be hundreds of kilometers long and 10-20 km wide. Heavier hailfall is associated with the broadening of the hail path. The typical period of heavier hailfalls is about 5 minutes. Maximum hail-size is centrally located and decreases on both flanks. The patchiness of the hail pattern and the intermittency of point hailfalls are characteristic of most of the hailswaths. Occasional overlap of hailstreaks of short duration occurs. Average maximum length and width of hailstreaks in Illinois is 9.4 km and 1.7 km respectively.

The observations of sequence of hail and rain show that all three categories are possible; rain first, hail first, or rain and hail together.

The detailed information of hailstreaks and associated rainfall is helpful in understanding storm dynamics, cell development and its waxing and waning effect.

CHAPTER 4

PHYSICAL CHARACTERISTICS OF HAILSTONES

Over the last thirty years, the problem of hailstone growth has come under ever-increasing scrutiny. The observational field networks as well as laboratory experiments have supplied a mass of data on the physical characteristics of hailstones. Before summarizing the available data, however, we shall distinguish between the various forms of ice particles which are sometimes collectively called hail.

4.1 Classification of hail:

The following definitions and descriptions are quoted from the GLOSSARY OF METEOROLOGY, American Meteorological Society (1959).

4.1.1 Hail:

> "Precipitation in the form of balls or irregular lumps of ice. By convention, hail has a diameter of 5 mm or more, while smaller particles of similar origin, formerly called small hail, may be classified as either ice pellets or snow pellets."

4.1.2 Hailstone:

"A single unit of hail, ranging in size from that of a pea to that of a grapefruit (i.e., from less than 1/4 inch to more than 5 inches diameter). The spheroidal stones, the most common form, typically exhibit a layered interior structure resembling an onion, with alternate layers composed of glaze and rime."

4.1.3 Ice pellets:

"A type of precipitation consisting of transparent or translucent pellets of ice, 5 mm or less in diameter. They may be spherical, irregular, or (rarely) conical in shape. Ice pellets usually bounce when hitting hard ground, and make a sound upon impact.

"Now internationally recognized, ice pellets include two basically different types of precipitation, those which are known in the United States as (a) sleet, and (b) small hail. Thus, a two-part definition is given:

(a) Sleet or grains of ice: generally transparent, globular, solid grains of ice which have formed from the freezing of raindrops or the refreezing of largely melted snowflakes when falling through a below-freezing layer of air near the earth's surface.

(b) Small hail: generally translucent particles, consisting of snow pellets encased in a thin layer of ice. The ice layer may form either by the accretion of droplets upon the snow pellet, or by the melting and refreezing of the surface of the snow pellet."

4.1.4 Snow pellets:

"(Also called soft hail, graupel, tapioca snow) Precipitation consisting of white, opaque, approximately round (sometimes conical) ice particles having a snow-like structure, and about 2 to 5 mm in diameter. Snow pellets are crisp and easily crushed, differing in this respect from snow grains. They rebound when they fall on a hard surface, and often break up.

"In most cases, snow pellets fall in shower form, often before or together with snow, and chiefly on occasions when the surface temperature is at or slightly below 0°C. It is formed as a result of accretion

of supercooled droplets collected on what
is initially a falling ice crystal (probably
of the spatial aggregate type)."

4.2 Hailstone shapes:

Most natural hailstones are either roughly conical, spherical or ellipsoidal in shape. The spherical form is by far the most common, especially when the size is small or medium. Sometime irregular stones are found and they show jagged and odd shapes with many external irregularities or lobes, particularly as they grow to the largest sizes. Different shapes of hailstones are shown in Figs. 4-1 to 4-4.

Conical hail is one of the often observed forms of hailstones. It grows by accretion of cloud droplets on an initial ice crystal or a frozen droplet. The conical form develops due to the stable orientation of the falling hailstone, which prefers droplet accretion by the lower side, resulting in the base growing most rapidly. The stones are usually opaque near the apex, becoming transparent or clear toward the base. The ice is opaque when the accreted droplets freeze relatively quickly at the point of impact; slower freezing produces clear ice.

The observational evidence suggests that conical graupel results because of riming. The problem is to explain how the riming starts on the frozen droplets or ice crystals and how the riming proceeds. We have found a number of explanations offered

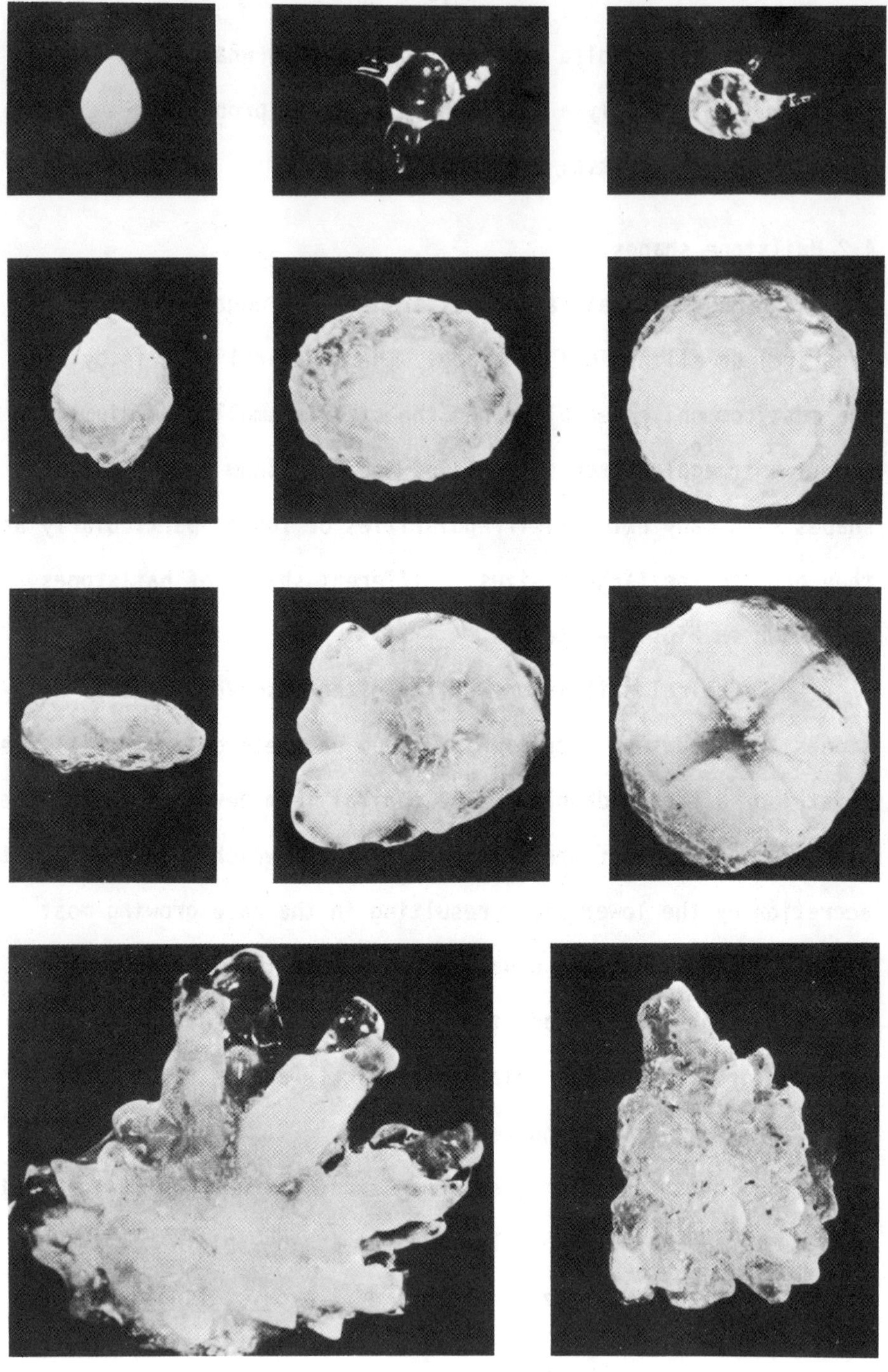

FIG. 4-1 Examples of hailstones of various shapes. (from Carte and Kidder, 1970)

(Reynolds 1876, Arenberg 1938, Knight and Knight 1973), which are speculative and do not seem convincing. An interesting suggestion is by Holroyd (1964). The conical graupel may be, according to him, aggregations of partially rimed needle crystals rimed further after aggregation. The mechanism of origin of conical graupel is important as they form a common stage in rain and hail formation, and hence this problem should receive more attention.

Relative frequencies of occurrence of different shapes of hailstones in various size groups are shown in Fig. 4-2. spheroidal hailstones outnumber other shapes in all size groups up to 5 cm. Cones exceeding 2.5 cm in diameter are rarely found although a few cones as large as 5 cm in diameter were observed. In the larger size groups, the ellipsoids, irregular shapes and 'apples' were more frequent.

The hailstones collected at Oak Ridge, Tennessee, by Briggs (1968), exhibited a variety of unusual shapes, some resembling starfish and daggers. The striking feature was the long and symmetrical spikes. Some of the stones had spiralled arms and others had lobes on the end of the spikes. Perhaps the most striking shapes were in the form of perfect crosses, or daggers.

Hailstones of very unusual shape and consisting almost entirely of clear ice (Figs. 4-3 and 4-4) have been reported recently by Mossop (1971). The fall of such unusual stones lasted only about a minute. It was followed by a light fall of stones of

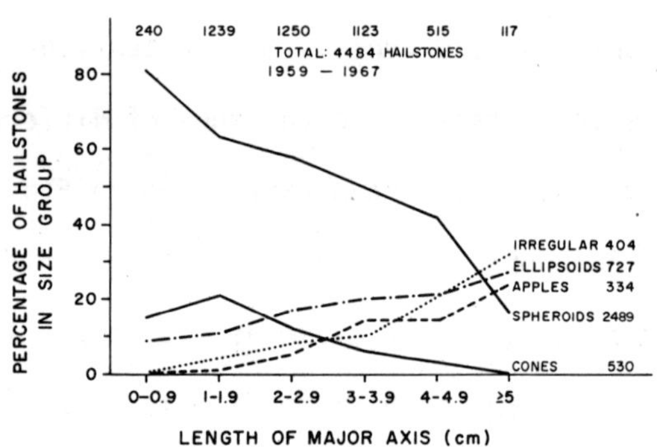

FIG. 4-2 Relative frequencies of occurrence of different shapes of hailstones in various size groups. (from Carte and Kidder, 1966)

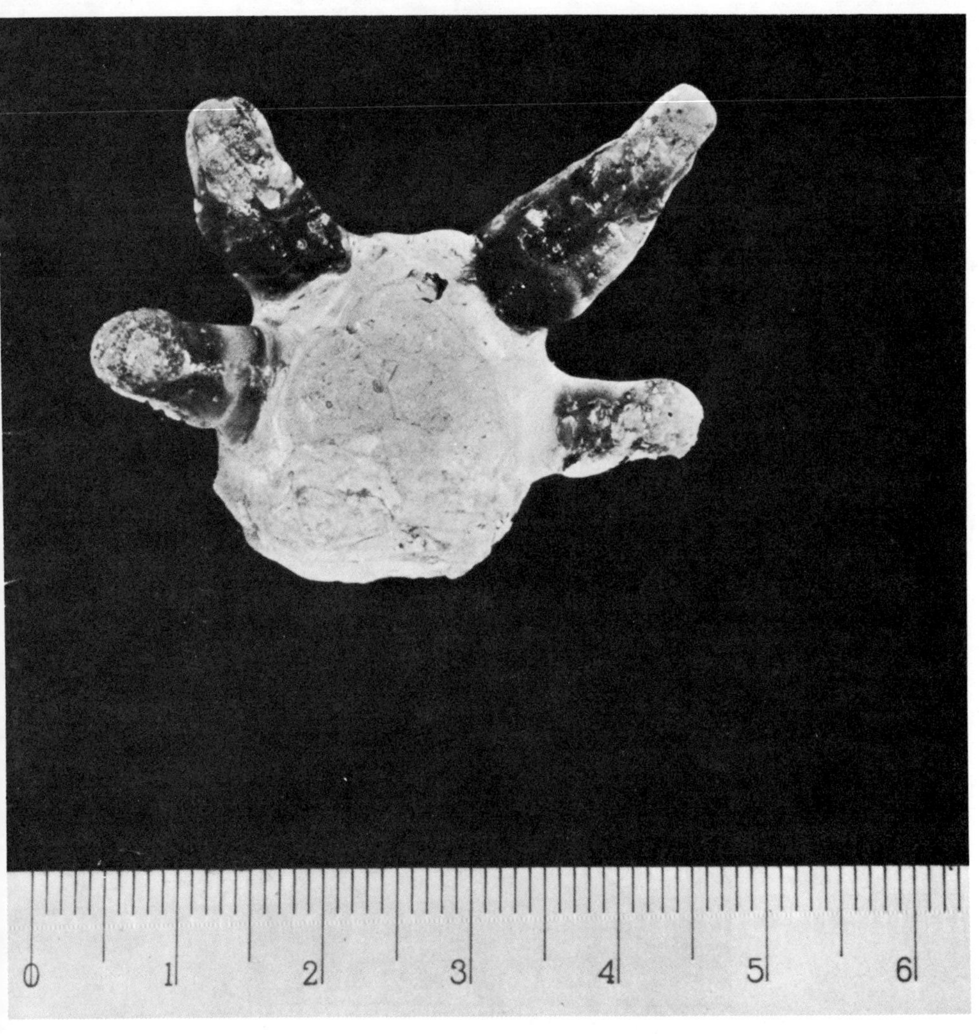

FIG. 4-3 Hailstone which fell at Sydney, Australia on 3 January 1971. The greatest dimension across the "horns" is 4.5 cm. Scale is in centimeters. (from Mossop, 1971)

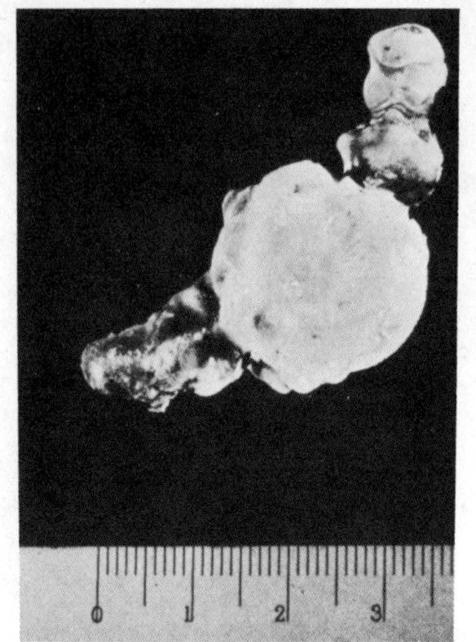

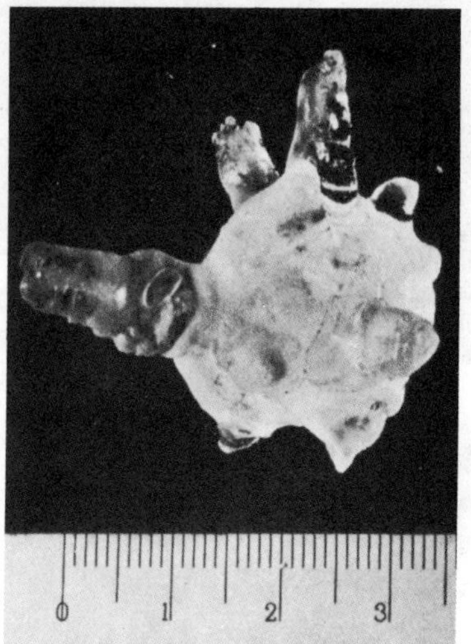

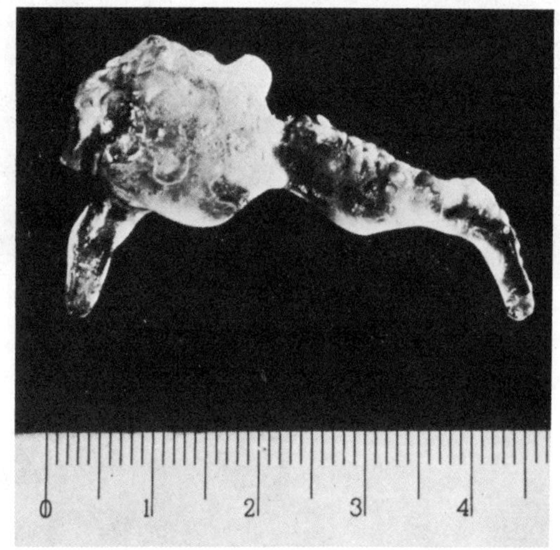

FIG. 4-4 Hailstones which fell at Sydney, Australia on 3 January 1971. The scale is in centimeters. (from Mossop, 1971)

more usual shape and then by rain. The embryos, 5 mm in diameter, of some of the stones were formed of milky ice. The clear layer of outer ice indicates that this growth was the result of the embryo collecting a large number of drops which could not freeze immediately on contact. The reader is referred to Sec. 5.9 for the details of a laboratory experiment confirming the wet growth of a lobe structure.

'Apple' shapes have been produced on growth centers in the laboratory by Kidder and Carte (1964). These embryos were rotated perpendicularly to an airstream in which supercooled water droplets were sprayed. The laboratory experiments confirmed that a single indentation results from a slight tilting of the axis from the normal to the airstream.

Oblate hailstones, including 'apple' shapes, were discussed in some detail by Browning and Beimers (1967). They have photographed thin sections of 90 large oblate hailstones. Throughout the growth period, the hailstones became increasingly oblate and this process continued during the melting of the hailstones. In extreme cases this process results in the development of pronounced indentations at one or both ends of the minor axis, forming 'apple' shaped hailstones.

4.3 Terminal velocities of hailstones:

The fallspeed or the terminal velocity of a hailstone is one of the parameters which determines its growth rate and maximum size which can be suspended in a given updraft. The terminal velocities of hailstones depend on the drag coefficient of spheres and other parameters discussed below (e.g., Bilham and Relf, 1937; Schumann, 1938; and Ludlam, 1958).

By equating the drag of a smooth sphere to its weight, the relation between terminal velocity and diameter has been deduced by Bilham and Relf (1937):

$$C_D \rho V^2 \pi d^2/4 = \sigma g \pi d^3/6 \qquad (4\text{-}1)$$

or

$$V^2 = 2\sigma g d/3\rho C_D \qquad (4\text{-}2)$$

where d, σ, V and C_D are respectively, the diameter, density, terminal velocity and the drag coefficient of the hailstone, ρ is the air density, and g is the acceleration due to gravity.

The empirically determined relation between C_D and Reynolds number (Re)

$$C_D = f(Re) = f(Vd/\nu) \qquad (4\text{-}3)$$

was used with equation (4-2) to compute the terminal velocity V in terms of d, ρ, σ and the kinematic viscosity ν of the air (Fig. 4-5). However, Bilham and Relf used a different definition of drag coefficient from that generally used today. This gave $C_D \simeq 0.25$ for spheres at a low Reynolds number, whereas now its

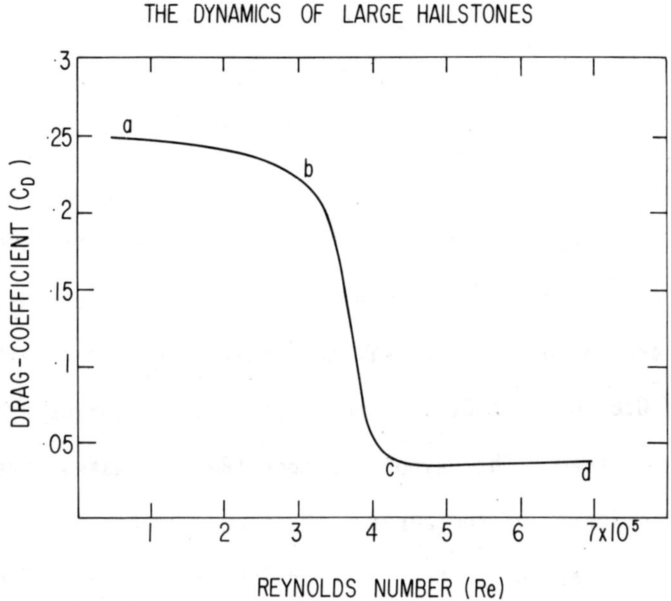

FIG. 4-5 Relation between drag coefficient (C_D) and Reynolds number (Re). (from Bilham and Relf, 1937)

generally accepted value is $C_D \approx 0.5$ (Macklin and Ludlam, 1961).

List (1959a) has determined the drag coefficients of four hailstones having masses between 18 and 32 g. For roughly spherical stones, the drag coefficients were 0.6, while for oblate stones with axis ratios 1 : 0.8 : 0.6, the values were nearer to 0.7.

Plots of diameter versus terminal velocity, applicable at the ground level, are shown in Fig. 4-6. The terminal velocity values are calculated for hailstones with specific gravity values of 0.4, 0.6, 0.8 and 0.915. The shapes of the curves are found to take the S-form. The Reynolds number (Re) increases steadily along the length of the curve. The stable branches AB and CD in Fig. 4-6 correspond to the flat portions ab and cd of Fig. 4-8 respectively. For values of Re between 3×10^5 and 4×10^5, the value of C_D falls rapidly due to the change in pattern of the "wake" and though d decreases, the value of (d/C_D) and, therefore V, increases. Thus, the back-bending portions BC of the curves result, which represent unstable conditions. Bilham and Relf were of the opinion that the maximum diameter of a spherical hailstone would be about 13 cm and its maximum weight about 0.7 kg, as the hailstones do not enter the upper regime of the terminal velocity.

Terminal velocity cannot be determined from C_D alone; an appropriate cross section is also needed. Bilham and Relf assumed that hailstones were smooth spheres. However, the natural

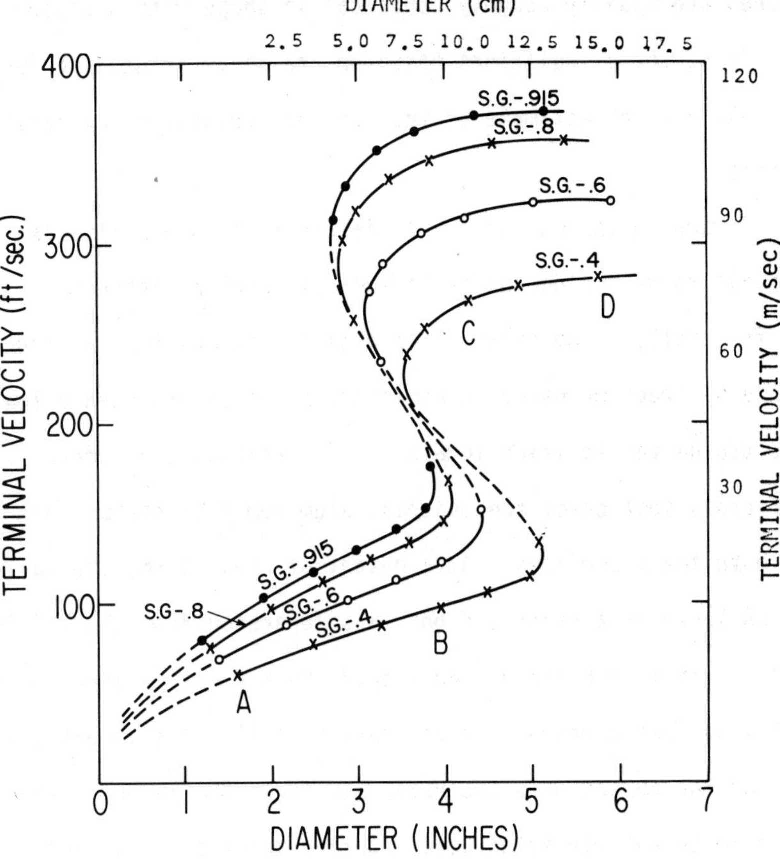

FIG. 4-6 Terminal velocity versus diameter for spherical hailstones having specific gravities 0.915, 0.8, 0.6 and 0.4 [ground level]. (from Bilham and Relf, 1937)

hailstones are usually roughly spherical in shape with protuberances. Thus, the calculations based on the above assumption furnished only a rough estimate of the terminal velocities of natural hailstones.

Macklin and Ludlam (1961) describe field experiments carried out to determine the fallspeeds of model hailstones during free fall, using radar. Ice objects resembling hailstones were made by freezing water in hemispherical cups and then allowing the two halves to stick together. To increase the backscatter cross section of the objects, aluminum foil strips were frozen onto their surfaces. At a height of about 2 km, the hailstone, enclosed in a cardboard box, was separated from the lifting balloon. A smaller balloon then tipped the box upside down and the hailstone was ejected. It was easy to follow the ascent and descent of the object once the radar had found the target. The range, azimuth and elevation were recorded automatically every five seconds. A plot of the height of each stone against time was made with the help of radar data. The accuracy with which the fallspeeds could be determined was $\pm$ 1 m sec^{-1} in most instances.

Figure 4-7 shows the plot of the measured fallspeed versus mass of the stone. The values of the radius of the sphere of equal mass are also shown. If the error is greater than $\pm$ 1 m sec^{-1}, the range of values is indicated. The curve

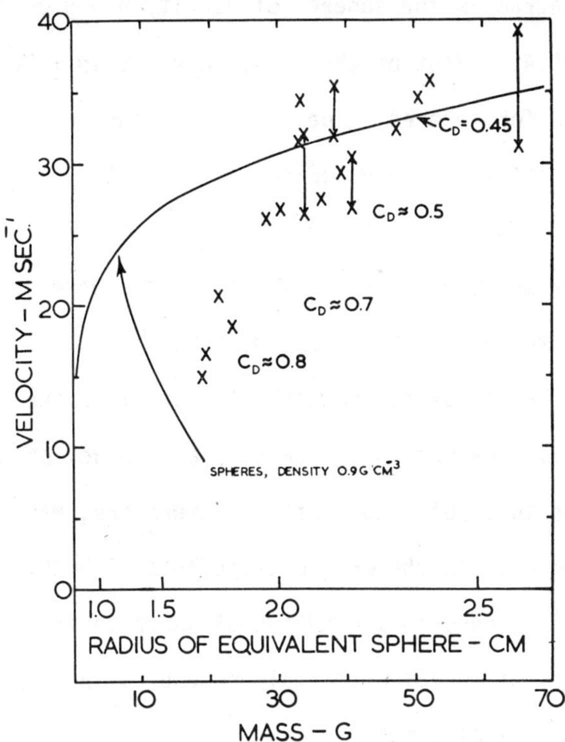

FIG. 4-7 The free-fall speed of a hailstone as a function of its mass. A scale showing the radius of the sphere of equal mass is also given. The curve for spheres is based on a mean drag coefficient of 0.45 and an air density of 1.15×10^{-3} g cm^{-3}, the latter value being appropriate to the present experimental conditions. Drag coefficients, estimated from the data, are indicated. (from Macklin and Ludlam, 1961)

in the diagram is for spheres of density 0.9 g cm^{-3} and C_D equal to 0.45. Most of the values are within $\pm$ 20 percent of the values for spheres. The scatter in the velocities may be due to slightly different orientations of the hailstones as they fell. The calculated value of drag coefficient for oblate stones was about 0.8, and their fallspeeds were considerably less than those of spheres of the same mass.

As it was not possible to determine the orientation of the hailstone during its fall, the accuracy of this method is limited to about $\pm$ 25 percent. Therefore, Macklin and Ludlam decided to measure the drag coefficients of both model and natural hailstones in a wind tunnel under controlled conditions. For details of the experimental procedure, the reader may refer to their original paper.

The drag coefficient is given by

$$C_D = 2D/\rho_a V^2 A \tag{4-4}$$

where ρ_a is the density of the air, V is the air speed, A is the cross sectional area normal to the airstream and D is the drag force. The value of $1/2\ \rho_a V^2$ was calculated directly from the manometer measurements, since

$$V^2 = 2\rho_w gh/\rho_a \tag{4-5}$$

where h is the pressure head, and ρ_w is the density of water.

Their data indicate that the drag coefficients of hailstones range from about 0.45 to about 0.8, depending mainly on three factors, namely, the shape, the Reynolds number, and the roughness of the surface of the hailstone. The values obtained for rough spherical stones were greater than those for smooth spheres for which the mean value was 0.45. In Fig. 4-8 the scheme proposed by Ludlam and Macklin for determining the drag coefficients of hailstones is shown. However, individual particles with surface irregularities may have drag coefficients differing appreciably from those plotted in Fig. 4- 8. Macklin and Ludlam's data indicate that the drag coefficients for hailstones, both of regular and irregular form, are considerably lower than those envisaged by Weickmann (1953), who suggested values in the range 1.0 to 2.0. In addition the data show that asymmetrical large stones ($d > 1$ cm) with mean value for the drag coefficient of 0.6, may have fallspeeds 15 to 20 percent less than those of spheres of the same mass.

Willis, Browning and Atlas (1964) made simultaneous measurements of the radar cross sections and fallspeeds of ice spheres ($d > 5$ cm) falling in free air from an altitude of 6 km. The radar had the operating wavelength of 5.47 cm. Their results indicate that large and dry hailstones have a drag coefficient as low as 0.24 to 0.3. But, this value nearly doubles when their surfaces become wet.

Young and Browning (1967) and Bailey and Macklin (1968) confirm that the drag coefficients of artificial hailstones of nearly spherical shape drop quite sharply when Re exceeds 10^5. However, this is not the case for lobed hailstones of larger sizes.

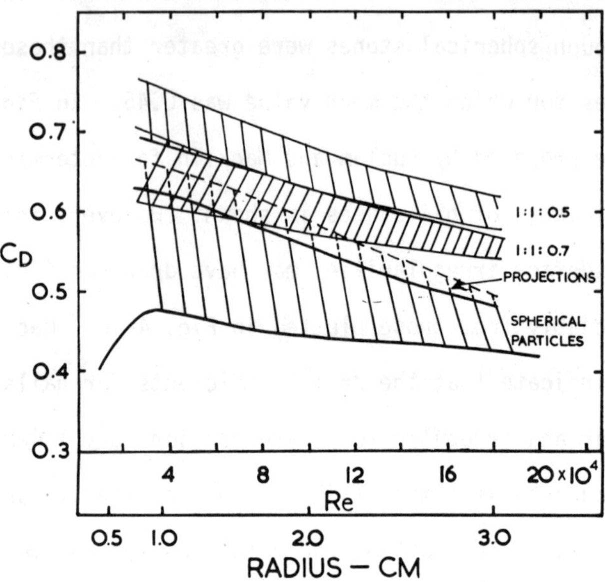

FIG. 4-8 Proposed scheme for determining the drag coefficients of hailstones, with an approximate scale of hailstone radius obtained by assuming C_D = 0.6, T = -10°C and p = 600 mb. (from Macklin and Ludlam, 1961)

Carte (1968) has reported the measurements of fallspeeds of polyethylene bodies of various shapes and masses while they were dropped down a 1.5 km vertical mine shaft. The results are shown in Fig. 4-9. The oblong and knobbly bodies marked (f), (g) and (h) fell appreciably slower than the other ones. There were difficulties in making these measurements as the falling bodies came in contact with the walls and other obstructions in the mine shaft. However, the measurements do show the effects of various shapes and masses on the fallspeeds, and these are in good agreement with the values reported by other investigators.

It must be noted that since the fallspeed depends on the air density, its value measured in a wind tunnel on the ground is less than the actual value at high levels.

Recently, aerodynamic drags of replicas of non-spheroidal hailstones were determined by Roos and Carte (1973) in a horizontal wind tunnel, and falling behavior was investigated in a vertical wind tunnel and a water column. Free fall experiments were also conducted indoors and from a helicopter. The empirical formula

$$V_T = Kd^{1/2} = 14.0\ m^{1/6} \tag{4-6}$$

where V_T is the terminal velocity in meters per second, d the diameter expressed in cm, and m mass in grams, yields velocities within $\pm$ 6% of measured values, for various non-spherical shapes, with K = 12.4. K = 15 is reasonable for spheroidal bodies.

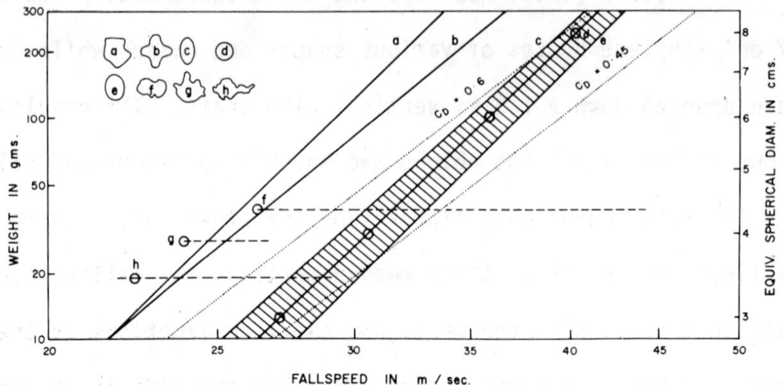

FIG. 4-9 Fallspeeds of polyethylene bodies of various shapes and masses. Dotted lines were calculated for spheres with drag coefficients of 0.45 and 0.6. Horizontal broken lines show the spread of fallspeeds as found by wind tunnel tests on replicas of three irregular hailstones, while the points are from the mine shaft experiments. All results apply to objects of density 0.9 g cm^{-3} falling in still air of density 0.001 g cm^{-3}. (from Carte, 1968)

Equation (4-6) becomes

$$V_T = 0.44\, m^{1/6}\, \rho a^{-1/2} \qquad (4\text{-}7)$$

when the effect of air density (in gcm^{-3}) on fallspeed is taken into account.

4.4 Hailstone sizes:

Figure 4-10(a), (b) and (c) shows histograms of hailstone sizes reported in the United States and Canada as summarized by Douglas (1963). The three distributions are similar in form and indicate 0.5 to 2.5 cm as the most frequent size range whereas 2.5 to 6.0 cm size range is rather rare. Frequency distributions of size of most common hailstones, as reported by Carte (1964) in South Africa and by Castet (1969) in France (Fig. 4-10(d) and (e) respectively) agree with the findings in the United States and Canada. It must be pointed out, however, that there are certain difficulties in making such comparisons between these histograms. For example, the Alberta data came from observed size distributions of many hail samples, while the remainder of the data came from observer reports of the largest or most common hailstones. In the case of South Africa, the histogram applies to the data from one exceptional storm.

Carte and Kidder (1966) report that smaller hailstones collected in simple mesh baskets undergo rapid melting when exposed to the heavy rain usually associated with hail. Moreover, despite appeals to cooperative observers, falls of small

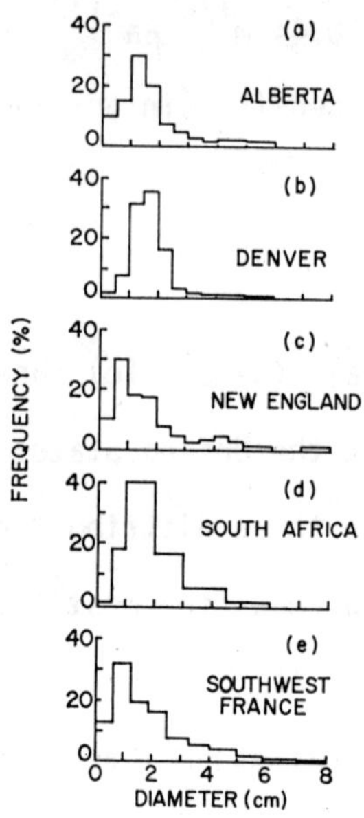

FIG. 4-10 Frequency distribution of hail by size in (a) Alberta, (b) Denver, (c) New England [after Douglas, 1963], (d) South Africa [after Carte, 1964], (e) Southwest France [after Castet, 1969]. (from Gokhale and Rao, 1969a)

hailstones pass unrecorded more frequently than the spectacular ones. The histograms therefore may not show very true relative frequencies of occurrences of smaller size groups.

One of the large hailstones reported from the USA which fell at Potter in Western Nebraska on July 6, 1928, had a diameter of 13.7 cm and weighed 0.7 kg. Recently it has been reported (Knight and Knight, 1971, and Roos, 1972) that a bigger hailstone fell at Coffeyville in Southeastern Kansas on September 3, 1970. The measurements of this hailstone were: 44 cm circumference and 0.77 kg in weight. Stones weighing 3.8 kg and 2.3 kg have been reported in Hyderabad, India (Hariharan, 1950), but these might have been aggregates of two or more hailstones.

4.5 Hailstone density:

The results of many measurements of density of medium and large size hailstones are now available in the literature (Steyn, 1950; List, 1958; Vittori and Di Caporiacco, 1959; Macklin et al., 1960; and Mossop and Kidder, 1961), which indicate values in the range 0.87 to 0.92 g cm^{-3}. The former correspond to an air content of about 5.4 percent by volume and the latter to pure ice.

Macklin et al. (1960) determined the densities of stones by immersing them in a series of chilled liquids. These were various mixtures of paraffin oil and carbon tetrachloride. Their specific gravities ranged from 0.850 to 0.916 in steps of 0.010. A calibrated hydrometer was used to determine these values to

$\pm$ 0.0005. By observing the velocity of fall or rise of the stones in the liquids, the densities were estimated within $\pm$ 0.003 g cm^{-3} This is a useful method only if the densities are high and there is no significant penetration of the liquid into the stones. Such penetration would be readily detected as the air bubbles would escape. Macklin and his associates also report density measurements of various layers of hailstones: clear ice has a higher density than opaque ice in which trapped air bubbles give rise to opacity.

List (1958) gives density of 0.5 to 0.7 g cm^{-3} for graupel particles of diameter up to 6 mm collected from summer showers at the Weissfluhjoch Observatory (2,665 meters above sea level). Some graupel observed by Nakaya (1954) and Magono and Nakamura (1965) had density as low as 0.2 g cm^{-3}.

4.6 Hardness of hailstones:

Weickmann (1953) reported observations of liquid water within natural hailstones. Vittori (1960) and Donaldson et al. (1960) observed hailstones which were sufficiently spongy to splash upon hitting the ground. Also, Carte (1963) and Browning et al. (1968) have reported mushy or soft hailstones which were composed of a mixture of ice and unfrozen water.

Calorimetric measurements of the liquid water content of hailstones collected at different locations have been reported by Gitlin et al. (1968). Fifty-seven percent of the samples collected in Kenya, Africa contained no water and these were hard hailstones. The remaining samples contained 4.2 percent water. The hailstones from Colorado and South Dakota showed water content variation in the range 0.4 percent to 14.6 percent.

Carte and Basson (1970) report useful information on soft hailstones in spite of the subjectivity of the data and the effect of melting. Those hailstones designated as soft may include mushy as well as graupel (low density rime), and those classified as hard might have water trapped in a solid shell of ice. The frequency distribution of soft hailstones and the details of the size effect are shown in Fig. 4-11. Twenty-two percent of the reports included some soft hailstones, and they were commonly encountered when only small hailstones fell. A seasonal effect is evident from the reports, namely, that soft hailstones are more likely to occur in winter than in summer.

It seems that soft hail is much more frequent in Alberta, Canada, than in the Transvaal (Summers, 1968). Forty-seven percent of the reports indicated the presence of soft hail. However, individual storms do show wide variations in soft hail frequency. Recent report by Federer and Waldvogel (1975) on hail fall from a Swiss multicell storm indicates that most stones were hard ice and not spongy.

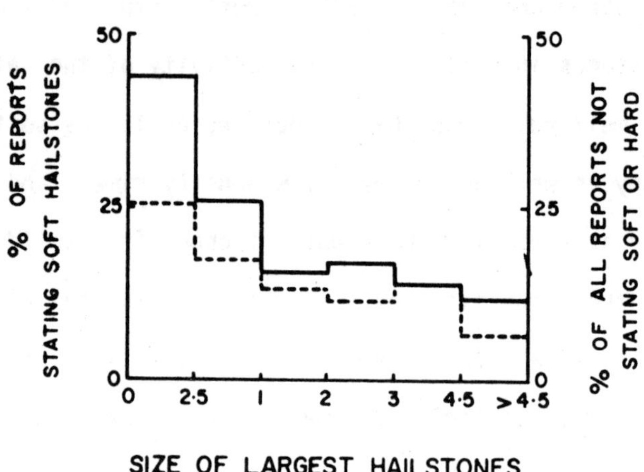

FIG. 4-11 Frequency distribution of soft hailstones [upper histogram]. Percentage of reports for which hardness or softness was not stated [dashed histogram]. (from Carte and Basson, 1970)

CHAPTER 5

INTERNAL STRUCTURE OF HAILSTONES

An individual graupel particle or a hailstone forms, as discussed earlier, by the accretion of a large number of supercooled droplets. The structure of the accreted ice on natural or artificial hailstones has been investigated in the laboratory by cutting a hailstone into thin sections and studying them under a microscope with transmitted and polarized light. With this technique the size and arrangement of individual ice crystals, the orientation of the larger crystals and the size and distribution of the air bubbles in the ice can be determined. Artificial hailstones are grown in wind tunnels and their structure is related to the ambient growth conditions, such as temperature, air velocity, liquid water content and droplet size distribution.

About 15 years ago three groups, one at the Swiss Federal Institute for Snow and Avalanche Research, another at the National Physical Research Laboratory in Pretoria, South Africa, and the third at the Imperial College of Science and Technology in London, started the studies of the internal structure of artificial as well as natural hailstones. Their pioneering work of gathering data on the internal structure reflected on the complexity of the subject. Groups in the USA and Canada are now continuing similar studies.

5.1 The Swiss Hail Tunnel:

The first hail tunnel at the Swiss Federal Institute for Snow and Avalanche Research was designed and built to grow hailstones artificially in the laboratory (List, 1960a). It was assumed that comparative observations of the artificial stones with natural hailstones would be helpful in understanding the conditions in which hail forms in nature. The basic construction of the tunnel is shown in Fig. 5-1. For details regarding different parts, the reader is referred to the original article by List. The different parameters controlled were air temperature, humidity, air speed, impurities in the air, droplet size and nucleating substances that were added to the moist air. The test section in which an ice particle was supported could be illuminated and observed through plexiglass windows as well as explored by various probes.

A wide variety of experiments on the growth of hail embryos was carried out in the wind tunnel and their structure was correlated with the different parameters stated above. The experiments with this first hail tunnel were useful in understanding some of the aspects of the growth of hailstones, especially the formation of spongy ice and its role in hail formation (List, 1959b, 1960a, 1961a; Macklin and Ryan, 1962).

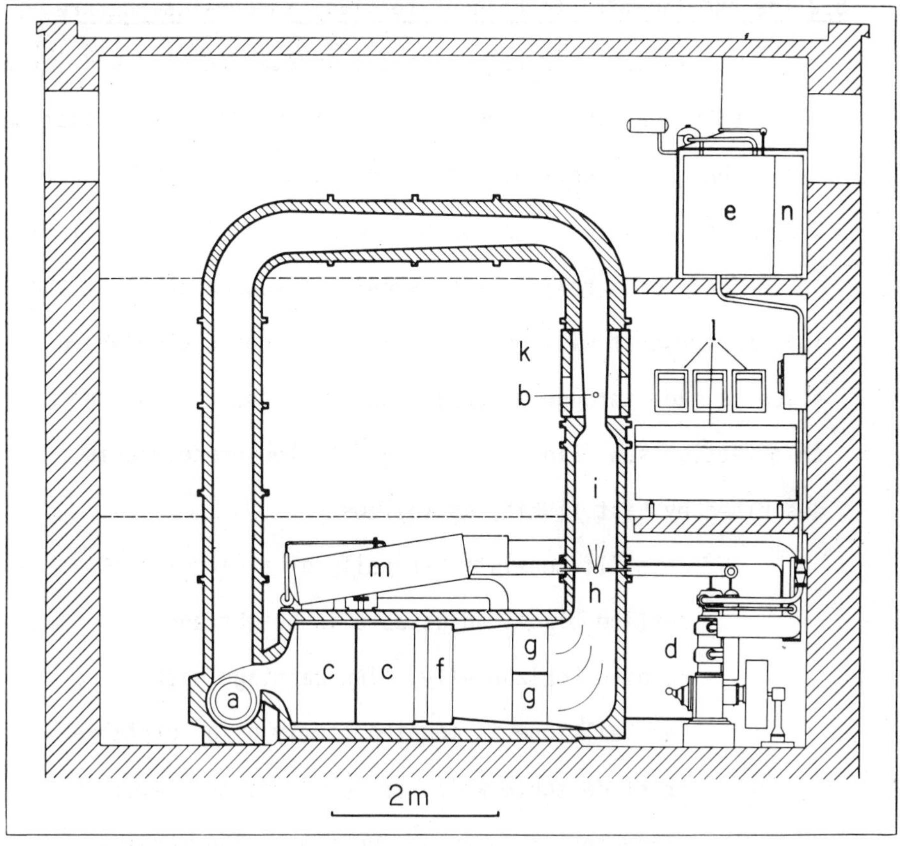

FIG. 5-1 General diagram of the hail tunnel; (a) blower; (b) test object; (c) air-cooler or vaporizer; (d) refrigerating compressor; (e) ammonia condenser; (f) heater; (g) electrostatic filter; (h) point where humidity and ice-forming nuclei are injected; (i) section where the flow achieves homogeneity; (k) measuring section; (l) control panel; (m) liquid separator; (n) alcohol-water cooler for cooling the cold compressor. (from List, 1960a)

5.2 The experimental techniques to study internal structure:

The air-bubble structure, crystal fabric, and layering of a hailstone can best be studied by looking at thin sections. The technique of obtaining very thin sections of ice down to 0.3 mm, or even less, was developed by M. de Quervain (1954). In the case of a hailstone it should be recognized that the information that will be available will depend upon where the slice is made through the hailstone with the help of the thin-section saw (Fig. 5-2). The detailed procedure is described by List (1961b) as follows:

> "It consists basically of a circular cutting blade, adjustable in height and running horizontally. The carriage with the hailstone is mounted firmly on a turntable or slide table which travels into the revolving blade and produces an incision. The machine represented in Fig. 5-2 and the cutting table shown separately in Fig. 5-3 have various advantages which should be briefly described:
>
> (a) The slide or mount with the hailstone is engaged by suction due to a vacuum between two mounting cylinders (Fig. 5-3). Excellent support is achieved by this method which cannot result in any distortion of the glass slide. This slide can be changed very quickly, particularly

FIG. 5-2 Thin-section saw, consisting of [1] cutting blade, [2] mechanical stage, [3] cutting table, [4] hailstone clamped in position. (from List, 1961b)

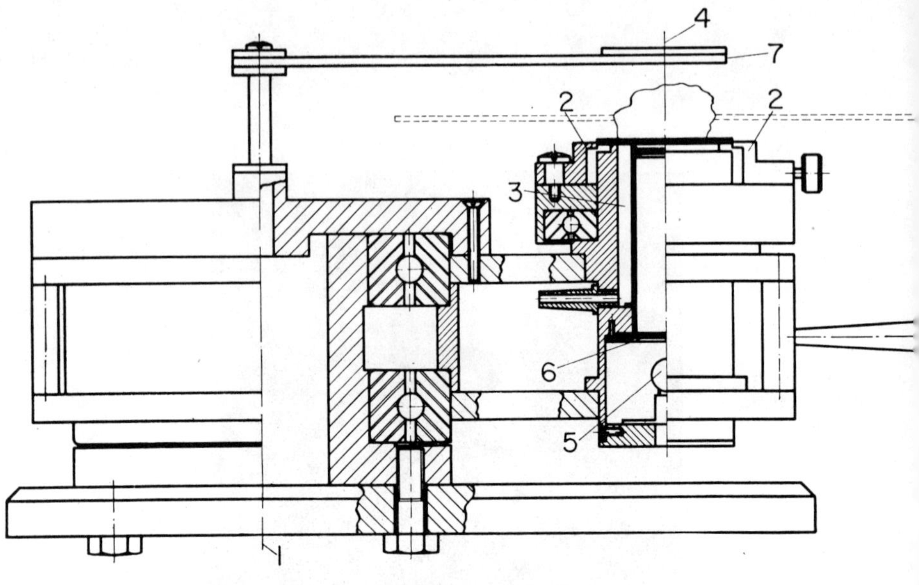

FIG. 5-3 Constitutive parts of the revolving cutting table: [1] revolving axis of the cutting table, [2] clamp for the slide, [3] cylinder ring with vacuum for fastening the slide by suction, [4] axis about which the slide clamp can be turned, [5] source of light, [6] polaroid filter I, [7] polaroid filter II [turned through 90 deg with regard to polaroid filter I]. (from List, 1961b)

as the bearing surfaces cannot get dirty. Apart from the suction mechanism, the slide is also lightly fixed mechanically.

(b) The mechanical fixture of the slide enables it to be turned about its own axis. This allows the hailstone to be cut from whichever side is most suitable.

(c) The circular velocity of the turntable is adjustable, and automatic. The hailstone does not therefore need to be inserted by hand into the cutting machine. A stop switch terminates cutting automatically.

(d) The thickness of the section can be measured during cutting if the analyser (a polaroid filter) is shifted over the axis of the specimen holder. Polarised light is generated along this axis which, when analysed gives the crystallographic picture of the ice section, and the colors which appear are suitable for the thickness.

(e) The thickness of the section can, however, also be checked directly by adjusting the vertical height of the cutting blade.

(f) The number of turns of the cutting blade is variable up to 120 rpm.

(g) By cooling the motor and drive mechanism of the saw, it is possible to prepare, at a temperature of even -10C, thin sections from loose ice structures.

(h) The most suitable blades to be used have large teeth and cut also on both sides.

"To prepare thin sections from ice of high density (i.e., with few air bubbles), the following steps have to be taken:

(a) A slight thaw is initiated on one side of the specimen hailstone by placing it on a heating plate at a temperature slightly above the freezing point of water until a sufficiently large flat surface has been created. This thawed surface must be located so that it is parallel to the main section - i.e., the section which we want to examine.

(b) This thawed surface is brought into contact with a slightly warmed glass slide. The hailstone must then be pressed against the glass so that the water remaining between them, on freezing, is not thicker than 0.02 mm. Since this intermediary layer produces no color effects in polarised light when it is as thin as this, the additional ice produces no change in the appearance of the crystallographic structure of the thin section itself. (For the examination of normal hailstones, the glass slide should measure 50 x 50 mm.)

(c) An intermediary stage is now necessary during which the water between the hailstone and the slide is frozen. This must be done slowly, as too rapid freezing - like a badly cleaned slide - will result in poor cohesion. It should be realized that the main pressure bears on this joint during cutting, so that it must not be weakened in any way.

(d) The slide with the hailstone on it is now fastened to the slide-table of the saw, and the height of the blade set so that the cut will go through the desired zone of the hailstone. (If the inner structure of the hailstone cannot be seen at the beginning, due to an opaque outer shell, then it is wisest to make the first cut at a position which will make it possible to see into the inner zones of the hailstone.)

(e) Cutting is done in such a way that the mechanical strain on the part of the hailstone which is frozen to the slide is kept at a minimum. This usually means that one saws from one side only as far as the middle of the hailstone, then stops the cutter and begins again from another side. Since, depending on the quality of the ice, small pieces can be torn out of the cut surface, it is advisable to set the blade about 0.3 mm lower so that the new surface can be made smooth.

(f) This surface represents one boundary of the zone from the hailstone which we wish to examine more closely. Another slide has now to be frozen onto this surface - the actual slide which will carry the thin section. The new slide is slightly warmed for this purpose, so that when it is placed on the cut surface sufficient ice will be melted to produce a faultless joint between the hailstone and the glass. If the slide is too warm, however, superfluous water will be created which will cause interference and have to be removed artificially later.

(g) A second pause is made here to give the liquid water time to freeze the hailstone onto the slide.

(h) The actual thin section itself is now produced, by first fixing the new slide onto the saw (the hailstone will now have been turned upside down from its former position) and then sawing through the hailstone at approximately only 1 mm above the thin-section slide. We then have two pieces: the rough thin section, and the remainder of the hailstone which is still stuck to the first slide. The rough thin section is now reduced in height by stages between 0.3 and 0.4 mm until it has the desired thickness. For color photographs of the arrangement of single ice crystals, a thickness of between 0.35 and 0.45 mm is to be recommended, while for a black and white evaluation an ice layer of between 0.25 and 0.35 mm is the best. It should be said that these figures are particularly suitable for photographs of the ice structure; for observing the air bubbles in the ice, and particularly for detecting their arrangement in "shells," it is far better to photograph thicker sections which can be reduced later.

(i) The thin-section glass slide is now cleaned and should be scraped with an erasing knife to remove all traces of superfluous ice or water which have been caused by thawing the hailstone onto the glass.

(j) To prevent vaporization of the thin section, it is placed on a glass plate (60 x 60 mm) covered in liquid paraffin, and the remaining space between the two pieces is then filled with this liquid.

"For clarity, the various stages in this process are represented in Fig. 5-4. We now have a thin section of a hailstone prepared ready for observation photography and further examination."

Some modification of this method is now in common use. A bandsaw rather than a circular saw is preferred and a microtome

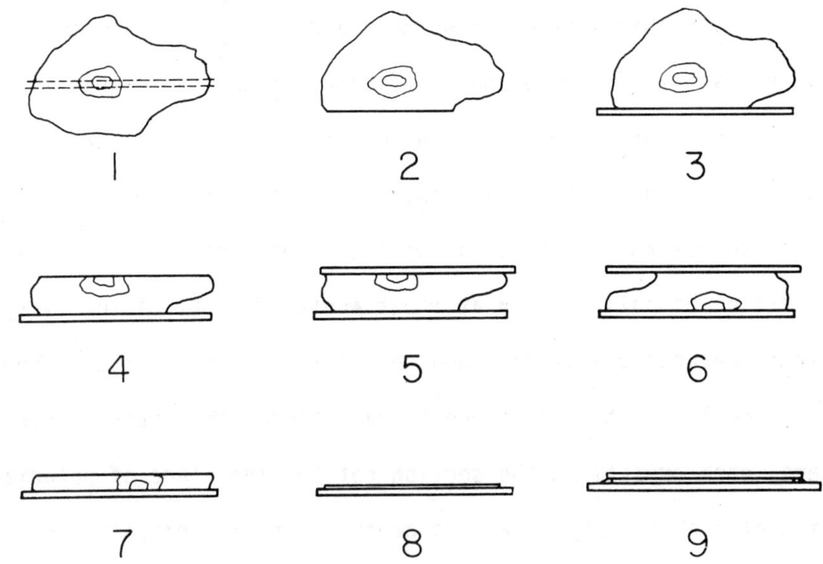

FIG. 5-4 Stages in the production of a thin section: [1] hailstone with intended place of incision, [2] preparing a flat surface by melting, [3] mounting on a warmed slide, and freezing, [4] fixing the slide onto the cutting table, cutting, and polishing, [5] freezing-on the thin section glass slide, [6] turning and mounting the thin-section slide, [7] cutting, [8] milling down to the desired thickness, [9] turning and laying onto a glass plate covered with liquid paraffin, filling of the intermediary space with liquid paraffin. (from List, 1961b)

is used for reducing the thick sections by melting.

A thin section shows only air bubbles and impurities in the hailstone if viewed under ordinary transmitted light. To examine the crystal structure, one has to view the section under polarized light. Two polarizing plates are used, one between the light source and the section and the other, orientated with its direction of polarization at right angles to that of the first plate, between the section and the microscope objective. When a thin section of the hailstone is put between the crossed polarizers, each crystal in the section rotates the plane of polarization of the light by an amount depending on the orientation of the crystal. Thus each differently orientated crystal appears in a different shade of gray, and the crystal textures are immediately visible. In additon, there are interference effects that give each crystal a different color. Black-and-white photographs taken with crossed polarizers give almost as much information.

Spectacular color photographs were taken using the above technique of mounting thin sections between crossed polaroids as shown in Fig. 5-5. (List, 1960b; Carte and Kidder, 1970; Knight and Knight, 1971)

5.3 Types of ice deposit:

Three types of ice deposit can be distinguished in the complex structures of hailstones:

FIG. 5-5 Thin sections cut from hailstones as seen between crossed polaroids. [Only half of the lower one was between the polaroids.] (from Carte and Kidder, 1970)

(a) Compaction is formed when the supercooled [illegible] spreads over the surface as a continuous [illegible] [illegible] the supercooled water [illegible]

[illegible]

(b) [illegible] when the [illegible] [illegible] with the environment is not [illegible] [illegible] [illegible] water. There exists [illegible] [illegible] [illegible]

[illegible]

[illegible] the number of air bubbles is large enough, it will have [illegible] [illegible] appearance. It [illegible] [illegible] [illegible] [illegible] are not common [illegible] [illegible] and [illegible]

(a) Compact ice is formed when the supercooled water spreads over the surface as a thin film before freezing. The transfer of heat between the deposited water and the environment is just sufficient to maintain the surface of the hailstone at 0°C. The ice is usually clear. The density of compact ice is in practice the density of pure ice, being close to 0.9 g cm^{-3}.

(b) Spongy ice results when the total heat exchange of a growing hailstone with the environment is not sufficient to freeze all the deposited water. There exists a critical amount of liquid water for given conditions, above which the surface becomes wet. The net result is that only a fraction of the collected water freezes to produce a framework of ice that aids retention of some of the unfrozen water, this mixture being maintained at 0°C. The density of spongy ice is between 0.9 and 1.0 g cm^{-3}. It is often quite transparent. However, if the number of air bubbles is large enough, it will have a slightly milky appearance. It should be pointed out that large spongy hailstones are not common. (Gitlin et al., 1968; Browning et al., 1968; and Goyer et al., 1968).

(c) Porous ice, which is sometimes referred to as rime ice, results when supercooled individual droplets freeze rapidly and produce a rimed structure. The ice deposit is then rapidly cooled down to a temperature between 0°C and the temperature of the environment. Air is trapped in the process, resulting in density as low as 0.2 g cm^{-3} in some cases. The trapped and tiny air bubbles scatter light and give the ice an opaque appearance; the lower the temperature of the deposit, the more opaque is the ice.

5.4 Hailstone embryos:

Examination of the internal structure of hailstones usually reveals an embryo of diameter less than 5 mm. But large hailstones with diameters greater than 2.5 cm often seem to have early growth centers ranging in size from 0.5 to 1.0 cm in diameter. Therefore, the part of an early growth center which constitutes an embryo is often difficult to determine.

Natural hailstone embryos have been studied by List and deQuervain (1953), List (1958, 1960b), Carte and Kidder (1966) and Knight and Knight (1970a).

The embryos are classified into four major categories:

(i) Conical graupel

(ii) Spherical, clear

(iii) Spherical, opaque

(iv) Irregular

These embryos are shown in Figs. 5-5, 5-7 and 5-9.

An embryo of a large hailstone may be either a graupel or a frozen raindrop. A graupel is a cone, or a sphere, or an ellipsoid of milky ice formed in the dry growth region due to accretion of small ice crystals and supercooled water droplets; it is a 'soft hail' particle, with maximum density of 0.8 g cm^{-3}. A frozen raindrop is either a clear or opaque ice particle, depending upon its freezing temperature; it is nearly spherical, consisting of large crystals randomly oriented. The density is about 0.9 g cm^{-3}. If the particle consists partially of liquid water and an outer shell which is frozen, the density may be as high as 0.99 g cm^{-3}. List (1960b) reported that 80 percent of all large hailstones collected in Switzerland had embryos of conica graupel. Macklin et al. (1960) studied the embryos and densities of 169 stones from a single storm and report that 90 percent of those stones had transparent centers. The center diameters ranged from 1.5 mm to 1 cm. It is probable that some were frozen raindrops and continued their growth in the wet stage.

The relative frequencies of various types of embryos in hailstones of different sizes have been reported by Carte and Kidder (1966); the curves are shown in Fig. 5- 6. The opaque spheroid was the type of embryo encountered most often in all size groups. However, 10 to 25 percent of the embryos were clear in different size groups. Carte and Kidder conclude from their studie that the shape of the growth center is evidently a dominant factor in determining the final shape of a hailstone. They found that

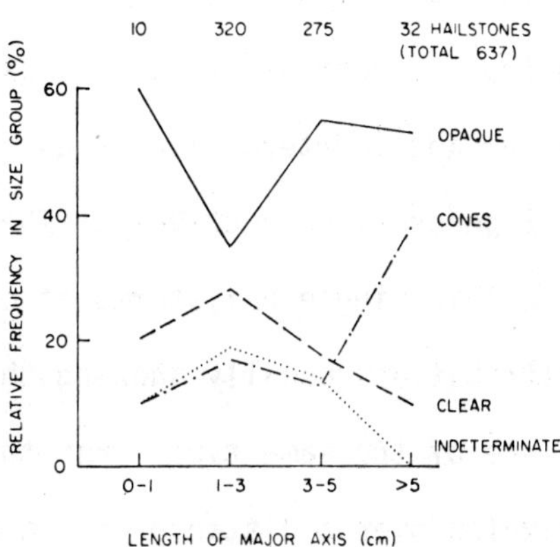

FIG. 5-6 Relative frequencies of various types of growth centers in hailstones of different sizes. (from Carte and Kidder, 1966)

the majority of the hailstones with conical embryos were ellipsoi whereas spheroidal growth embryos resulted in spheroidal hailston in seventy-five percent of the cases investigated.

In a sample of approximately 400 stones from 40 storms in Colorado, Knight and Knight (1970a) report the relative frequencies in four categories — conical, spherical (clear), spherical (opaque) and indeterminate — as 60 percent, 25 percent 10 percent and 5 percent, respectively. There is less variabilit in embryo form within single hailstorms at one location on the ground, the distributions usually showing that more than 80 perce of the embryos are of the same type. But the collections in one storm which traveled over a distance of 16 km showed a consist change in embryo type. Conical bubbly embryos were noted in the beginning, while spherical and usually clear embryos were observe toward the end.

Rogers (1971) has described an unusual embryo of an ellipsoidal stone, having a major axis length of about 3 cm and a minor axis about 1 cm. A thick section of this hailstone in Fig. 5-7 shows that its embryo is composed of two fused particle

5.5 Hailstone layers:

The internal structure of hailstones usually exhibits alternate layers of clear and relatively opaque ice beyond the embryo. Alessandro Volta was aware of the layered structure of hailstones as early as 1800 (Fig. 5-8). The majority of hailsto

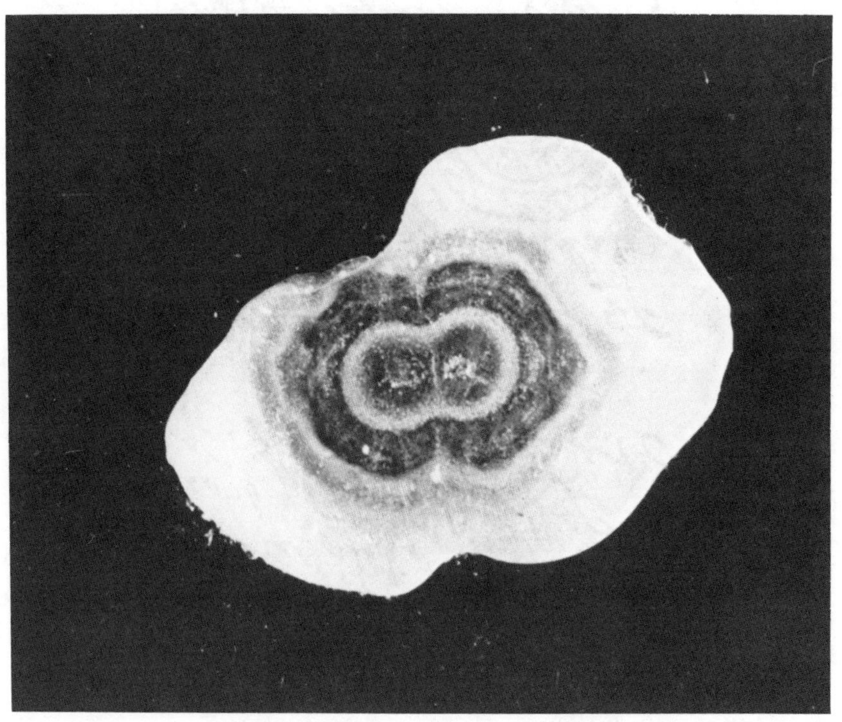

FIG. 5-7 Thick section of hailstone showing fused embryo. Major axis of hailstone is 2.6 cm. (from Rogers, 1971)

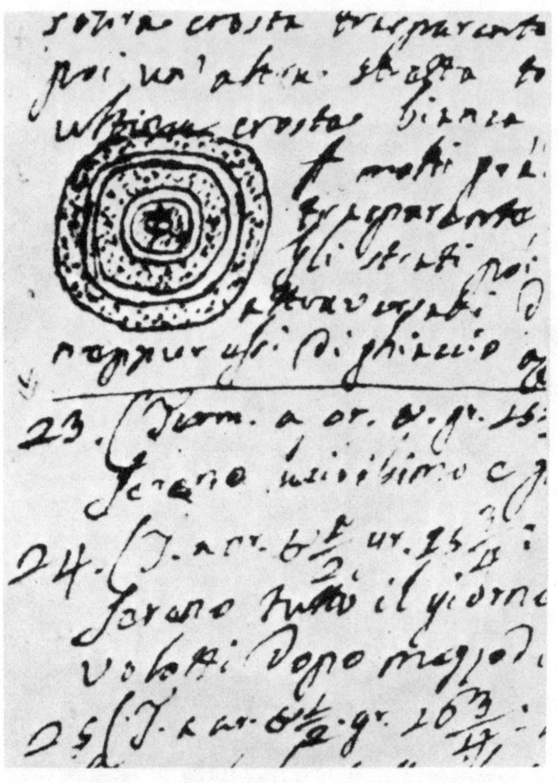

FIG. 5- 8 A sketch of a multi-layered hailstone made by Alessandro Volta in the year 1800. [This portion of an original Volta manuscript is reproduced by kind permission of the Instituto Lombardo Scienze e Lettere di Milano.] (from Carte and Kidder, 1970)

greater than 3 cm in diameter consist of either three or four layers (Fig. 5-9) but the number of layers can be as great as eight or even more. However, the layered structure which is shown in Fig. 5-9 is not found in all stones; some are composed entirely of either clear or opaque ice.

Mossop and Kidder (1961) studied fifty-five stones from a single storm in Australia, ranging in weight from 7 to 50 grams, and reported that six of them had more than four well-defined zones. However, the 4-zone structure was the commonest form as deduced by examination of thin slices in transmitted light, both polarized and unpolarized. About 40 percent of the stones had only three zones, the outermost clear layer being absent. A sketch of a large stone having seven complete zones and one incomplete zone is shown in Fig. 5-10.

The solubility of air in water increases as the temperature is lowered. Thus, the opacity of ice is determined by the amount of air dissolved in water; this dissolved air can be trapped in the form of air bubbles when the water freezes rapidly, giving rise to opacity.

A more meaningful criterion, however, for layering is the changes in crystal structure rather than in opacity. Where the crystal structure changes, opacity also changes in most cases; however, the opposite is not true. Carte and Kidder (1966) report the relative frequency of number of crystal layers as shown in

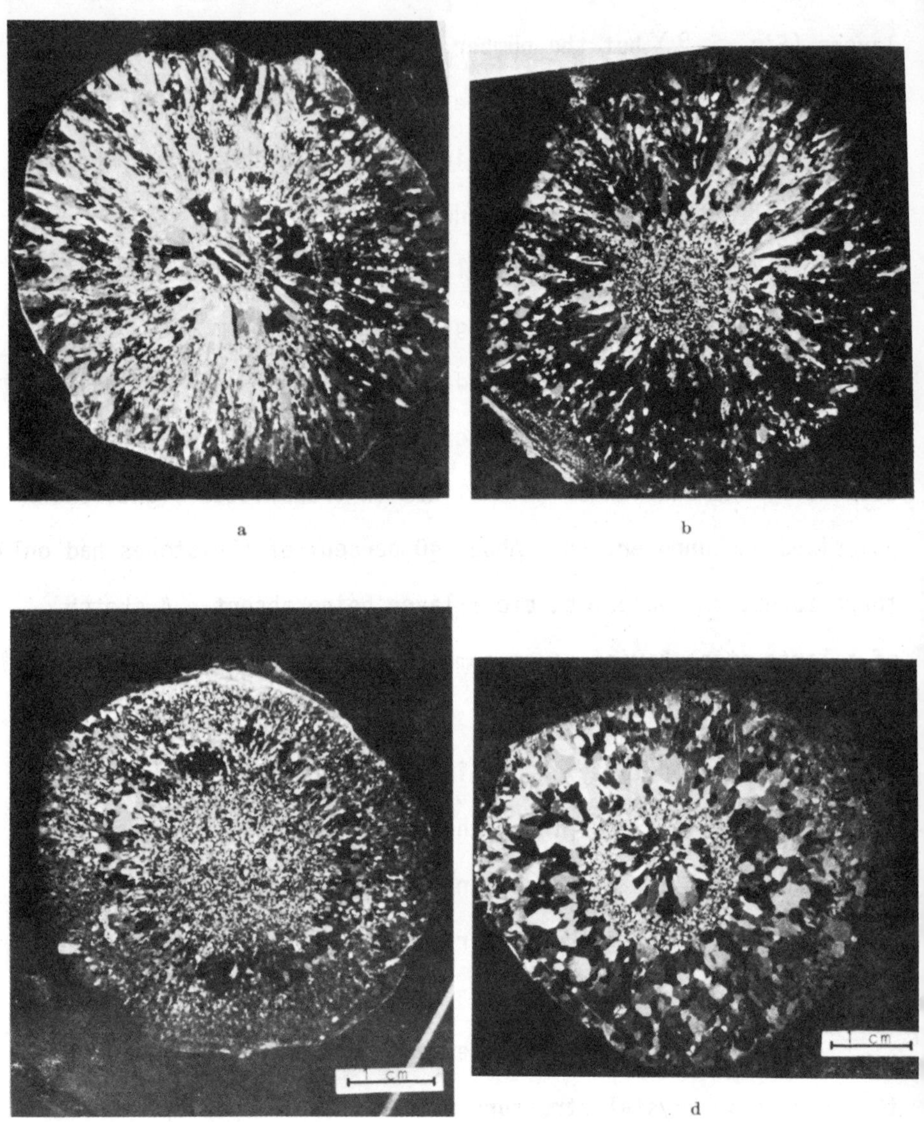

FIG. 5-9 Variations in hailstone cross sections. (from Schaefer, 1960)

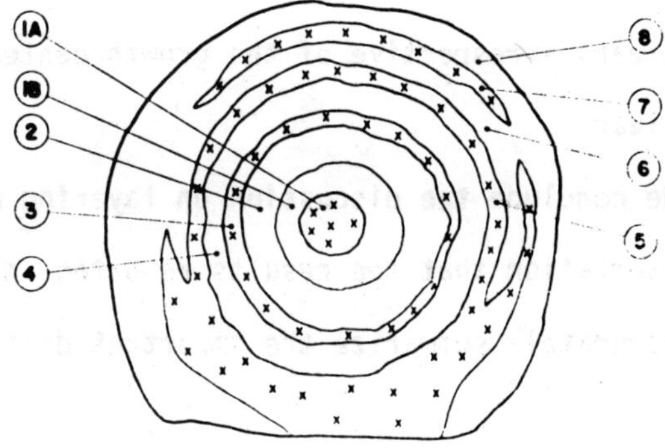

FIG. 5-10 Hailstone 252 - Diagram of structure of thin section. (from Mossop and Kidder, 1961)

Fig. 5-11. The number of layers most often found increases from 2 to 4 with the size of the hailstone, as expected. The greatest number of layers was eight in the biggest hailstones. Harrison and Beckwith (1951) report that all stones in a particular sample from a Denver storm had a clear outer layer. On the other hand, Carte and Kidder (1966) found stones with an opaque outer layer. The outermost layer is just as likely to be opaque as clear, irrespective of the growth center which may be opaque or clear.

We conclude the discussion on layering of hailstones with the observation that the results reported in the previous paragraph accurately summarize the important details on this subject.

5.6 Crystal size and shape:

The crystal (the grain is called a crystal in conformit with the general practice in meteorological literature) size and shape are striking parameters when ice accretions are observed between crossed polaroids. The studies by Carte and Kidder of thin sections of hailstones photographed between crossed polaroid (1970) indicate crystal dimensions in the size range 100 μ to 1 c The smallest crystals were all considerably larger than the cloud droplet sizes. It is suggested that some changes in crystal size might have occurred during recrystallization. The crystals may

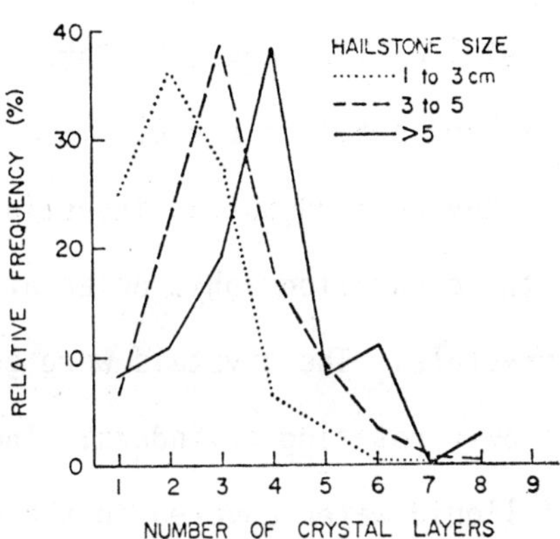

FIG. 5-11 Relative frequency against number of layers defined by changes in crystal structure for medium, large and extra-large hailstones. 673 hailstones were investigated. (from Carte and Kidder, 1966)

be wide or narrow and radially elongated. Many times they are complex, randomly oriented and intertwined. These characteristics are shown in Fig. 5-12.

A quantitative assessment of the sizes of the crystals is very complicated. The arbitrary distinction is made by Levi and Aufdermaur (1970) calling crystals large if $L > 2$ mm, medium if $2 > L > 0.5$ mm, and small if $L < 0.5$ mm, where L is the length of the crystal. They carried out an investigation in a wind tunnel to study the crystallographic orientation and the size of the individual crystals. The crystals were grown by the accretion of droplets on slowly rotating cylinders. The range of temperature and fractions of liquid water covered in the experiment are given in Fig. 5-13. The straight lines indicate approximate transition zones between large, medium and small crystals. Large crystals were noted when the air temperature was warmer than $\approx$ -20°C. Small or medium size crystals appeared below this temperature. From crystal size it may be roughly possible to decide whether accretion took place in an atmosphere warmer or colder than $\sim$ -20°C. For high to low wind speeds the transition shifts from near the colder limit AB to the warmer limit CD as shown in Fig. 5-13. The important conclusion of this study was that the parameter of greatest influence on crystal size was the ambient air temperature.

There is some minor agreement between the results

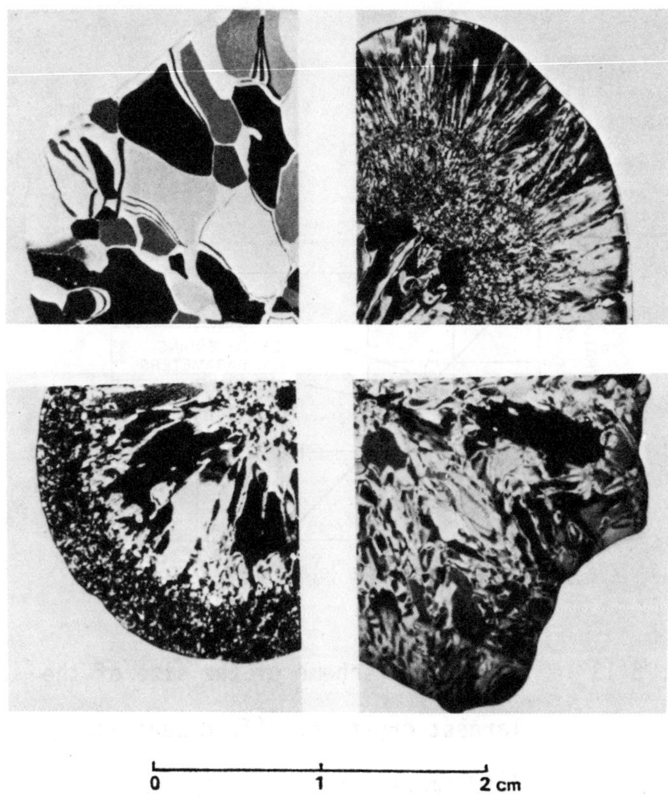

FIG. 5-12 Different sizes and shapes of crystals in thin sections from four hailstones. (from Carte and Kidder, 1970)

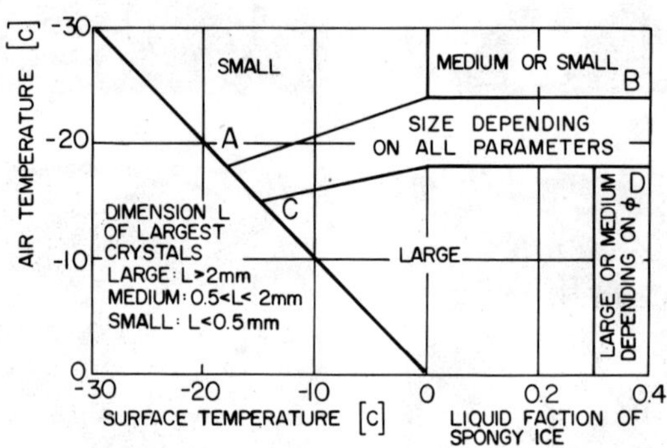

FIG. 5-13 Simplified scheme of the size of the largest crystals. (from Levi and Aufdermaur, 1970)

summarized in Fig. 5-13 and those published by Brownscombe and Hallett (1967). According to their results, the temperature of the growing substrate is much more important than any of the other variables.

The average surface area of crystals allows a separation of the opaque and transparent or semi-transparent zones according to the recent studies reported by List et al. (1970). Their data on one of the two hailstone samples are shown in Fig. 5-14. Crystal surface area equal to $5x10^{-3}$ cm^2 separates the two zones.

5.7 Crystal orientations:

Measurement of crystal orientations in hailstones may give some insight concerning the growth mechanism of the hailstone.

Higuchi (1958) described a method of determining the orientation of ice crystals in commercial ice by thermal etching. Painter and Schaefer (1960) developed a technique with which a permanent record of the crystalline structure can be made. Koenig (1962) combined these two techniques to obtain a permanent record of the crystalline structure as well as of the orientation of the individual crystals within a hailstone. The details of Koenig's method are quoted below:

> "The hailstones are cut and the surface polished by methods such as those described

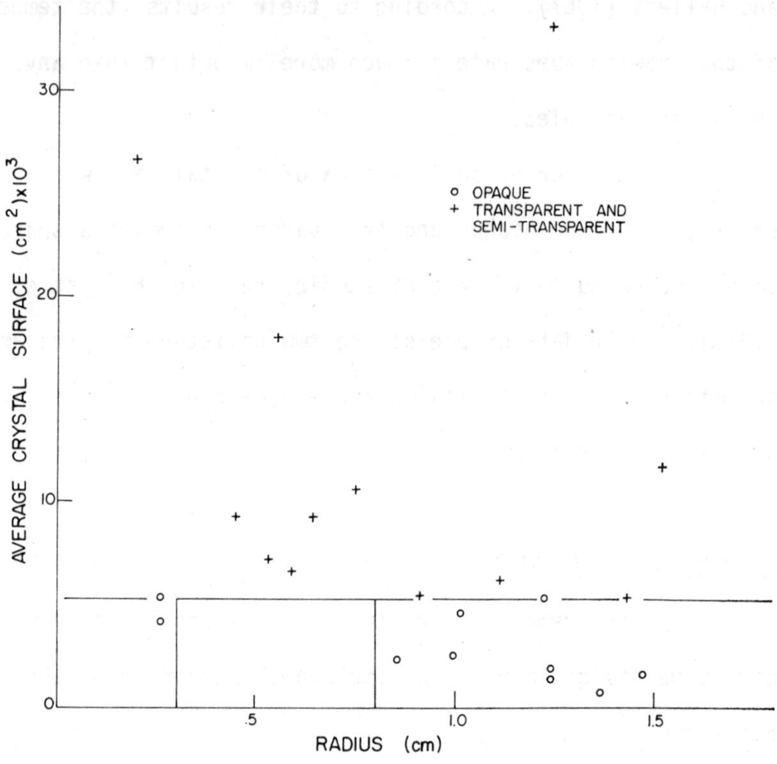

FIG. 5-14 Average crystal surfaces as a function of radius for Sample II. (from List, Cantin and Ferland, 1970)

by Painter and Schaefer (1960). In the example shown in this paper, a saber saw having a blade with coarse, wide-set teeth was used to cut the specimen. Waterproof silicon carbide paper and a cloth were used to polish the face on which the structure was to be revealed.

"The polished surface is then etched by placing it in an environment having a relative humidity lower than 100% with respect to the surface temperature of the ice section. In the present case this was often done by first warming the specimen by placing it about a foot above a small lamp in the cold chamber in which the processing was done and, after a few minutes, transferring the section to a cooler region of the freezer. As a result of the temperature gradient, the surface of the section was etched and resembled a metallurgical specimen prepared in order to reveal its grain structure.

"The etched surface is then covered with a 1/2 % solution of Formvar in chloroform which is allowed to harden. It is then reetched in a similar manner. The first etching process results in a more or less even removal of material across any given exposed crystal whereas the second etching step which is carried out beneath the plastic cover results in localized removal of material from the crystal surface. In this way etch figures are developed that have been shown by Higuchi (1958) to accurately indicate the specific orientation of the crystal on which they have formed. A permanent record of both the crystalline structure and its orientation now can be made by applying a generous coating of 3% Formvar in chloroform solution to the surface. The first plastic coating will become redissolved in the solvent before the hardening has been completed and, consequently, this film need not be removed. After the solvent has evaporated, the Formvar remains as an accurate negative replica of the surface of the hail section. The replica can then be removed, mounted between thin glass plates and examined conveniently with the aid of a microscope. The approximate orientation of the crystals may be easily determined by inspection of the replica. By means of well-known crystallographic procedures such as described by Higuchi (1958), the precise orientation of each crystal may be established (Fig. 5-15)."

The fabrics of several natural and artificial hailstones have been measured by Aufdermaur et al. (1963). They reported that the natural hailstone had a fabric with the c-axis preferentially radial. However, artificial spongy hailstones had a fabric with the c-axis preferentially tangential, indicating that the natural hailstones they studied did not grow spongy. Similar studies of orientation fabrics of natural hail by Knight and Knight (1968) confirmed these findings. List et al. (1970) have determined the crystallographic orientation distributions of ice crystals in natural hailstones. Their studies indicate that the average surface area of crystals and modal frequency of the c-axis distribution are sensitive indicators which may enable a separation of the transparent and opaque shells.

A 'dark cross' is observed when a thin section containing a complete ring of radially elongated crystals is viewed between crossed polaroids (Carte and Kidder, 1970; Levi and Aufdermaur, 1970). The cross remains fixed when the specimen is rotated as it is parallel to the planes of polarization. Such a cross usually allows one to easily detect a strongly radial orientation.

The study of the crystallographic orientation of

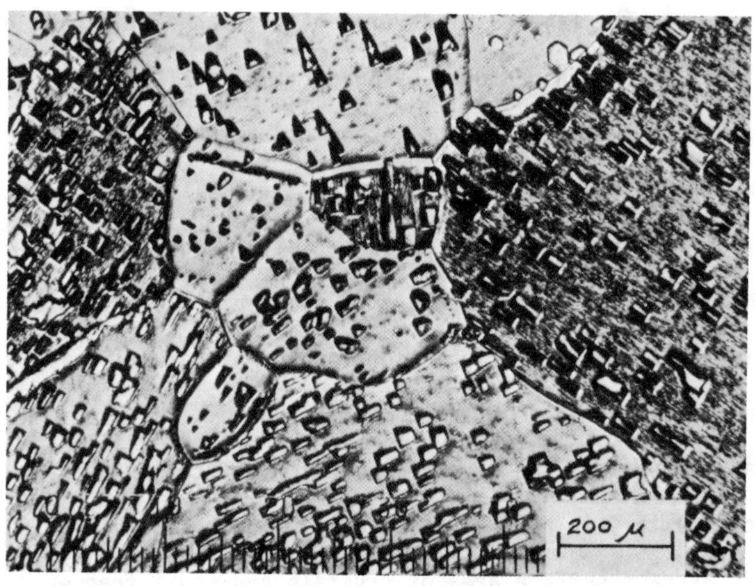

FIG. 5-15 Magnified view of replica which shows details of the central structure of the hailstone. (from Koenig, 1962)

crystals grown from accreted droplets on slowly rotating cylinders in a wind tunnel was carried out by Levi and Aufdermaur (1970) and Macklin and Rye (1974). Wet as well as dry growths were studied. The orientation of the c-axis was normal to the growing surface in dry growth, whereas it was parallel to the plane of the growing surface in wet growth. More detailed information on the environmental temperatures may be obtained by looking at individual crystals and their c-axis orientations as there are peaks in the distributions which may be broadly related to the temperature of the accreted droplets. However, this procedure is rather difficult and requires evaluation of replicas.

A different type of approach was taken by Brownscombe and Hallett (1967) and Macklin and Payne (1969). They examined the crystal fabric of the individual drop following the accretion process. Brownscombe and Hallett compared the accreted drop on a snow crystal to the accretion on a giant hailstone. However, the results obtained from these individual drop studies are difficult to compare with those obtained by other workers with wind tunnel studies, in which very large numbers of drops were accreting on the hailstone. The relation between size of new crystals in accreted ice to air temperature and surface temperature is represented by Fig. 5-16. When the air temperature is below a critical value, new crystal orientations form. This critical value is lower for smaller droplets. Large crystals are produced when a

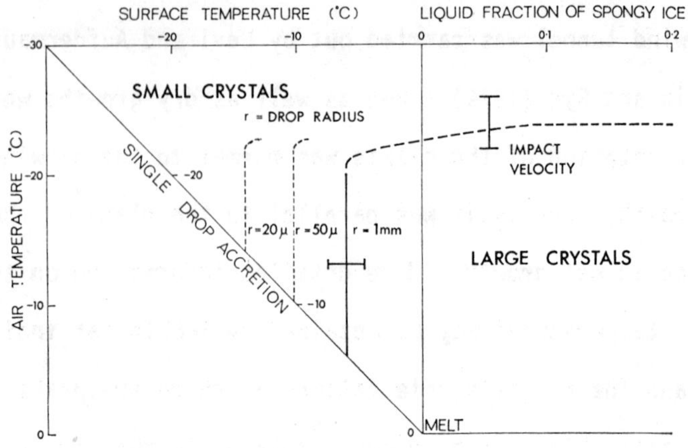

FIG. 5-16 Relation between air temperature, surface temperature and the formation of new crystals in accreted ice. Large crystals grow when a drop freezes entirely with the orientation of the ice on which it lands. The boundary between large and small crystals is displaced to lower temperature for small drops. At low air temperatures, new orientations may be formed at large impaction velocities. (from Brownscombe and Hallett, 1967)

drop freezes entirely with the orientation of the substrate ice. Supercooled drops were observed to retain the orientation of the substrate when they were impacted at low velocities, i.e., epitaxial growth. However, when they were impacting with large velocities during spongy growth, the air was trapped within broken dendrite mesh.

In conclusion, it may be said that such information on crystal orientations is difficult, time consuming and expensive to obtain. Hence, the technique has not been widely used. It is also difficult to make significant deductions from orientation data since the most important factors in preferred orientation development have not yet been definitely identified.

5.8 The air bubbles:

Dorsey (1940) and many others have made measurements of the solubility of air in water. Solubility increases with increase of pressure and decrease of temperature. The solubility of air in ice is about 10^{-3} times less than in water. Thus as water freezes, the air is rejected from solution at the growing interface. The formation of air bubbles released during the freezing of bulk water has been reported by Carte (1961); Carras and Macklin (1975). The concentrations and bubble sizes are dependent on number of factors such as rate of freezing, quantity of dissolved air, movement of the water and escape of bubbles. The relation between bubble radius and growth rate is shown in Fig. 5-17.

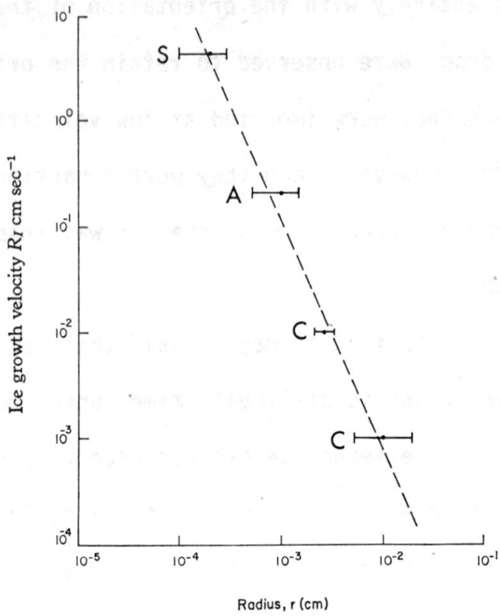

FIG. 5- 17 Experimental relation between bubble radius and growth velocity of ice.

C = Carte, 1961.

S = Freezing in supercooled liquid at -8°C.

A = Drop on aluminum plate at -20°C.

(from Brownscombe and Hallett, 1967)

(Brownscombe and Hallett, 1967). These experiments, though valuable toward understanding the physics of air bubble formation in supercooled bulk water, do not provide explanations for the complex distributions of air bubbles in hailstones.

Different patterns of air bubbles become apparent when sections of hailstones are examined in detail (List, 1958; Browning, 1966; Browning et al., 1968; Knight and Knight, 1968; and Carte and Kidder, 1966, 1970). Four such patterns are shown in Fig. 5-18. The bubbles are arranged sometimes in radial chains or in concentric shells. They may also be scattered at random throughout a layer. The sizes vary considerably from submicroscopic to as large as several millimeters. Shapes vary from spherical, cylindrical, and oval to highly complex ones.

Carte and Kidder photographed hailstone sections about 4 mm in thickness to study air inclusions. The sections were made to include the maximum diameter of the hailstones. The observations are summarized below:

(i) Chains of air bubbles in radial or curved lines were apparent in about twenty-five percent of the sections.

(ii) The lines of air bubbles may be obvious or ill-defined, numerous or few.

(iii) Bubble lines do not necessarily follow grain boundaries.

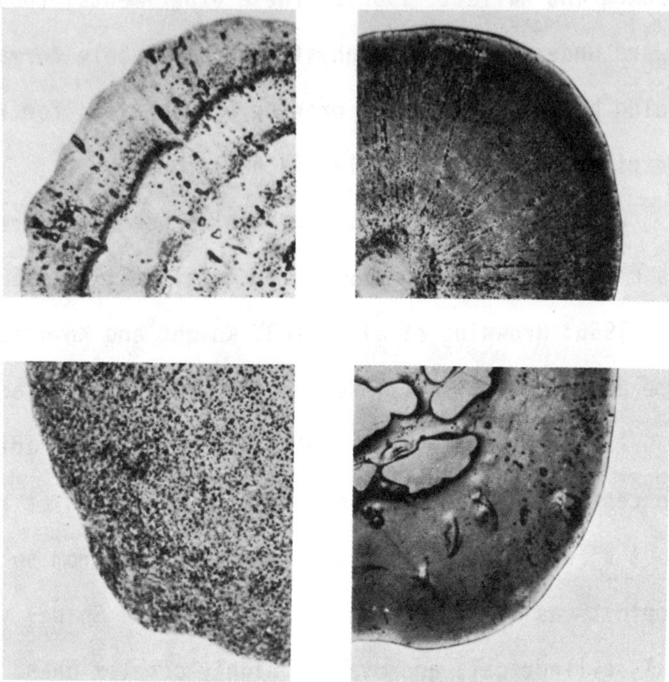

FIG. 5-18 Quarters of thin sections from four different hailstones showing different patterns of air bubbles.

Scale: 0 — 1 — 2 cm

(from Carte and Kidder, 1970)

(iv) Radial lines of air bubbles are often found near the sides of cones.

(v) The lines of bubbles are frequently found on the boundaries between lobes in sections of hailstones.

(vi) Narrow radial bands of air bubbles are seen in fairly clear, knobbly hailstones.

(vii) Air inclusions of millimetric dimensions are often to be seen in layers of transparent ice when thick sections are examined.

(viii) Elongated cavities are found in both opaque and transparent ice.

It seems, then, that the investigations during the past decade of the internal structure of natural and artificial hailstones have added many details to our knowledge as discussed in sections 5.6, 5.7 and 5.8. According to some, the structural studies would furnish qualitative information on the growth environments of hailstones; but this has not yet been achieved. The complexities of the problem are many. For example, the crystallographic orientation, size, shape and spacing of crystals in accretions of ice are dependent on a number of variables such as temperatures of the air and hailstone surface, the type of growth, updraft speed, liquid water content and the droplet size distribution. Furthermore, there is a danger that some recrystallization and structural changes may result in the processes of storage, cutting and polishing of the stones. The studies have also made apparent certain problems and ambiguities in the methods presently used for interpreting hailstone structures, resulting in some pessimism as regards to the usefulness of such studies of structure of hail as a diagnostic tool.

Whether or not these difficulties can be overcome is uncertain. Perhaps other techniques such as determination of isotopic ratios (refer to discussion in Chapter 9) may provide the necessary information on the growth environment of hailstones, such as liquid water content, temperature, height, etc.

5.9 Lobe structures:

Natural hailstones exhibit a wide variety of lobe structures of which two are the most common: the cusped type, and the "icicle" type. The cusped type has been described by Sarrica (1965) and others. The latter type has been discussed by Weickmann (1953), Knight and Knight (1970b) and Gokhale and Spengler (1973). The lobe structures can be classified on the basis of internal and external characteristics of large hailstones. They are extremely important and must be accounted for in any treatment of the growth of large hail.

The development of lobe structures seems to result when any protrusion from the surface grows faster than the surroundings. The protrusion thus becomes more and more marked.

5.9.1 Cusped lobes:

This type of lobe structure has been discussed by Sarrica (1965), Browning (1966), Bailey and Macklin (1968), and Knight and Knight (1968, 1970b). Figure 5-19 shows thin sections from large hailstones with cusped lobes.

The analyses of the internal bubble and crystalline structures of the hailstone sections with cusped lobes indicate that they are composed of a three-dimensional array of lobes which give the surface a convoluted appearance. These lobes are completely frozen, sometimes separated by regions of spongy ice characterized by radial lines of bubbles. The lobes are strongly convex so that successive growth layers become convoluted or scalloped as viewed in thin sections. The lobe-like structures may commence quite early in the growth of stones and have been found in ones as small as 1 cm in diameter, though diameters of 2 to 3 cm are more common. Browning (1966) has suggested that surface irregularities so produced affect the airflow around the stones and hence the drag and heat transfer coefficients. Thus, the hailstones can grow to a large size without becoming excessively spongy.

Bailey and Macklin (1968) were able to reproduce in their laboratory lobe structures very well. They used a rather simple but effective icing tunnel, which was housed in a cold room. The hailstones were freely supported in a vertical airstream. Its updraft speed could reach more than 40 m sec^{-1}. Experiments were conducted in the temperature range -5 to -30°C.

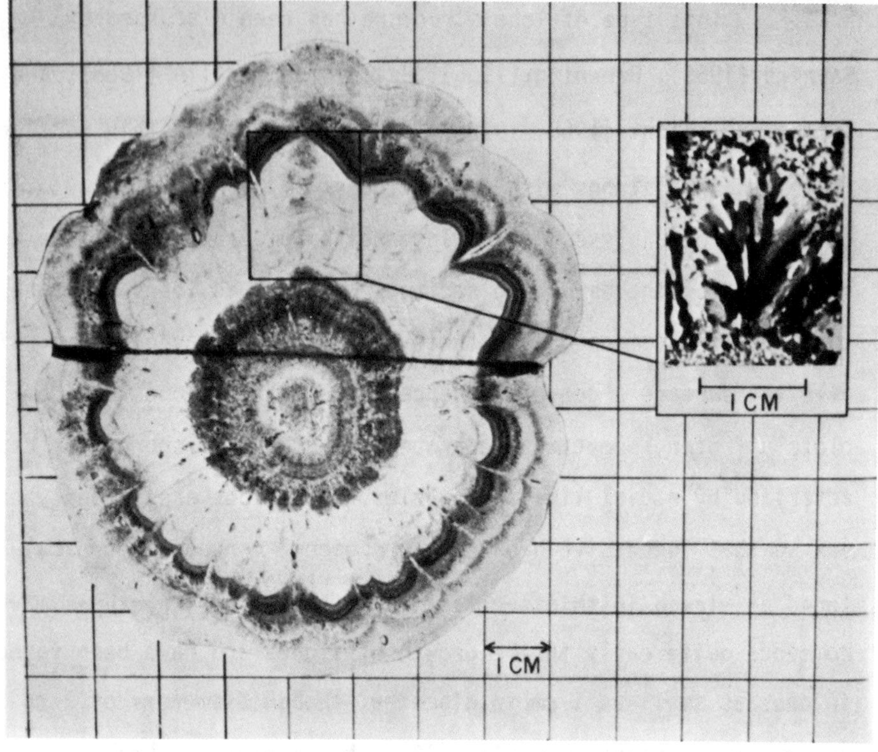

Fig. 5-19 Thin section close to the growth center of a hailstone showing the bubble structure; regions of clear ice appear white and milky ice appears black. The insert in this figure shows the crystal fabric of one of the lobes. (from Browning, 1966)

Water droplets were sprayed into the tunnel with various spray nozzles. The drop sizes varied between 29 μ and 240 μ in diameter. Most of the droplets had cooled to within a fraction of a degree of the ambient temperature by collision time. The artificial hailstones shown in Fig. 5-20 were grown in this manner, and

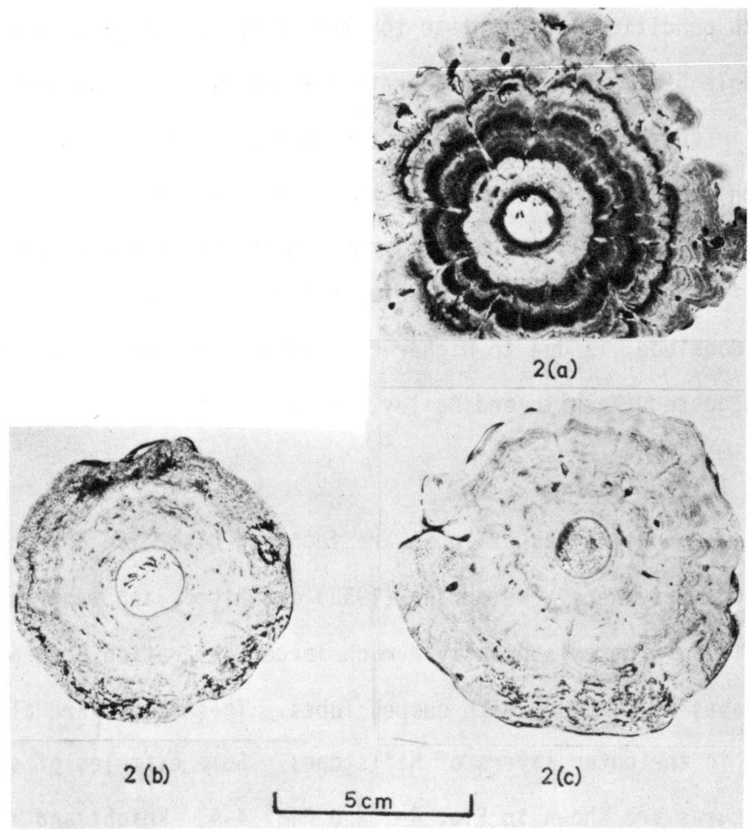

FIG. 5-2 0 Sections showing the bubble and crystalline structure of artificial hailstones. [The growth conditions appropriate to each stone are given in Table 5-1.] (from Bailey and Macklin, 1968)

are very similar to natural ones and show distinct lobes. The growth conditions appropriate to each stone in Fig. 5-20 are shown in Table 5-1. From their laboratory experiments Bailey and Macklin conclude that the lobes are most pronounced when the growth is taking place near the wet limit, and when the accreted droplets are small. Lobes are produced in dry growth as well and deep fissures sometimes form between them. The formation of lobes, they conclude, is due to higher collection efficiency as previously suggested (Macklin and Bailey, 1966).

5.9.2 Icicle lobes:

These types of lobes are found in clear ice and they are rounded everywhere. Weickmann (1953) described each protrusion as an icicle. There is usually a much larger separation between icicle lobes as compared with cusped lobes. These lobes are always found in the outer layers of hailstones. Some examples of such structures are shown in Fig. 4-1 and Fig. 4-4. Knight and Knight (1970b), from their microstructural studies of such stones, observed that more extreme icicle lobes grew wet due to freezing of thin liquid layers on the surface and were only slightly spongy.

5.9.2.1 Artificial growth of "icicle" lobe structure of a hailstone in a wind tunnel:

It is probable that these lobes form in the same manner as icicles form, with water flowing over the ice structures and

TABLE 5-1

The growth conditions for the hailstone sections

shown in Fig. 5-20. (from Bailey and Macklin, 1968)

Fig.	Average radius (cm)	Growth time, t (min)	Times of removal from tunnel (min)	Ambient temperature T_a (°C)	Median volume diameter of droplets, d_m (μ)	Average effective liquid water concentration (g m^{-3})	Comment
(a)	5·4	26	11 and 18	-22 to -19	29	4·3	Dry, but tending towards wet growth limit as the stone grew larger
(b)	3·5	27	13	-13 to -10	240	2·9	Spongy
(c)	5·1	34·5	25·5	-12 to -9	57	3·0	Spongy on lower side

freezing at the tips of projections. In an effort to shed more light on the factors important in the growth of these structures, experiments were performed in the author's laboratory (Gokhale and Spengler, 1973). A ping-pong ball with a thin ice coating served as the original simulated hailstone. A length of wire ran through the ball and was fastened at both ends. This arrangeme allowed the ball to rotate about 180° but restricted its movement up and down. The entire suspension was placed in a large vertical wind tunnel (Spengler and Gokhale, 1970; Spengler, 1971a) [Fig. 5-21] which could suspend many hundreds of water drops simultaneous thus simulating <u>some</u> of the conditions during the growth of a natural hailstone in a cloud. Water drops were produced by a vibrostaltic pump which pulsed a stream of water into 120 drops per second. The mean drop size was about 5 mm diameter, and the updraft speed was set at around 10 m sec^{-1}. The ambient air temperature was -17°C. The suspended drops above the hailstone had sufficient time to cool to at least -5°C.

Detailed observation of the hailstone and freely suspend hydrometeors above it was made possible by the use of high magnification and high-speed photography (Spengler and Gokhale, 1971). Film strips were taken using a 16 mm camera with close-up attachments and using framing rates between 200 and 1000 frames per second, thus facilitating the study of the high speed water drop impactions.

Initially, the hailstone on a horizontal wire was

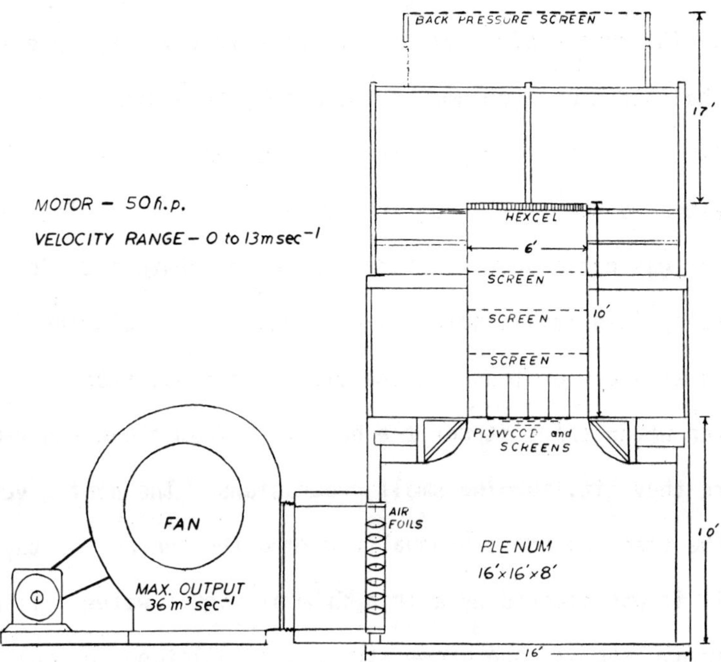

FIG. 5-21 Vertical wind tunnel - Department of Atmospheric Science, State University of New York at Albany. (from Spengler and Gokhale, 1972)

assumed to be at the same temperature, -17°C, as the ambient air. The supercooled water drops were freely suspended above the hailstone as a column of two meters in depth. They gradually drifted into the turbulent wake of the hailstone and were quickly accelerated downward, usually striking the stone. This wake effect was noticeable as high as 2 meters above it. Some drops broke up into many smaller sized drops after rupturing in the turbulent wake or touching the side of the hailstone. A few frozen water drops struck the hailstone from above and remained where they hit, forming small protrusions. The heat given off by the freezing drops gradually warmed the surface of the hailstone until it was covered by a thin film of liquid water. It began to accumulate as more drops struck the hailstone and collected mostly around the equatorial region. As the water froze there, we observed small, smooth bumps gradually forming in that region. The water flowing over these bumps gradually formed a larger protrusion, 0.5 cm long, in the equatorial region. The weight of the ice in this region was then sufficient so that the hailstone rotated on its axis a great deal in the updraft, and the protrusion oscillated between a downward and a horizontal position.

The protrusion or icicle type lobe then grew faster as more water accumulated, ran down the length of the icicle, and finally was shed at its tip (see Fig. 5-22). The water usually broke off in tiny droplets which shot upward in the wind tunnel. Gradually, the icicle grew so large (about 2 cm in length and around 1 cm across) that it always pointed downward. However other smaller protrusions started to grow in all directions.

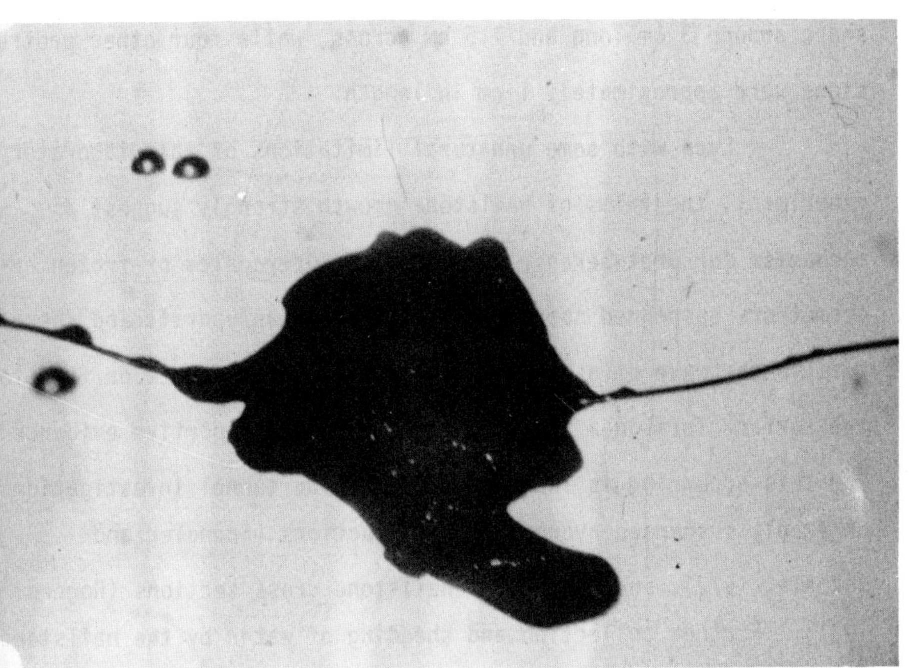

FIG. 5- 22 Artificial growth of "icicle" lobe structure of a hailstone in a wind tunnel. (from Gokhale and Spengler, 1973)

After about 10 minutes of total growth time, the hailstone was approximately 5 cm in diameter as compared with a starting size of 4 cm. In addition, the largest icicle growth had achieved a bent shape around 3 cm long and 1.5 cm across, while four other protrusions were approximately 1 cm in length.

Even with some unnatural limitations of this laboratory experiment, the films of hailstone growth strongly suggest a mechanism for protuberance formations. Supercooled or frozen hydrometeors suspended above the hailstone in an updraft and therefore in the wake of a growing stone contact and freeze or fuse to the surface forming a protuberance. Further supportive evidence for this occurring is found in earlier wind tunnel investigations on freely suspended hydrometeor interactions (Spengler and Gokhale, 1972), and in natural hailstone cross sections (Rogers, 1971). Further collection and shedding of water by the hailstone causes preferential growth of the protuberances. There are three reasons for this preferential growth. First, it has been demonstrated that protuberances have their own stagnation points (Bailey and Macklin, 1968; Spengler, 1971b), and therefore are locally efficient collectors of supercooled water. Secondly, since the entire stone is free to rotate, the unfrozen liquid water on the surface is forced to the region of highest angular velocity which would be the protuberance. The films show water being shed from the growing protuberances as expected. Thirdly, as Browning (1966) has noted, hailstone lobes significantly enhance the efficiency of heat loss from the surface.

CHAPTER 6

RADAR ANALYSIS OF HAILSTORMS

Undoubtedly radar is one of the most powerful tools of the meteorologist in severe storm research, with aircraft, radiosonde and surface observations also playing major roles. The introduction of weather radar has led to tremendous development in snow and rainfall observations. It reveals the history and structure of the three dimensional precipitation patterns. A number of studies for over two decades have been attempted with radar to make unequivocal detection of hail and the determination of hail size. Though a positive means of making such a detection has not been discovered, the latest developments in the technique of using dual-wavelength radar, as discussed in section 6.4, do show some promise.

Progress in radar meteorology during the period 1951-1957 has been comprehensively reviewed by Marshall and Gordon (1957). Battan (1959) is the author of the first textbook devoted entirely to this subject; he later authored a popular book on the same subject (Battan, 1962). In addition, there are two excellent reviews of progress in radar meteorology by Atlas (1963 and 1964). From 1955 to 1965 there was a sudden resurgence of interest in the study of scattering by hydrometeors. Data processing and display techniques were also considerably improved during this period. An entirely new technique for velocity measurements, namely Doppler radar, came into effect in the early 1960's. This instrument has tremendous possibilities in the determination of quantitative particle size distributions, vertical motion measurements, and large scale mapping of the horizontal wind field. In the pages to follow, we plan to summarize the use of radar as applied exclusively to hailstorm studies.

6.1 Characteristic forms of hail echoes:

The United Air Lines radar evaluation group (Harrison and Post, 1954) made airborne observations during 1953 to 1955 of severe thunderstorms in the Denver area with the 5.5 cm radar. The frequency of hailfall is very high in this region. They obtained more than 400 radar pictures of hail-producing thunderstorms, and correlated them with hail-network reports. Their experiences indicated that the characteristic forms which distinguish hail from rain echoes were (i) the "pointing finger" 1 to 5 miles in length, (ii) the "hooked finger" of the same dimensions, and (iii) blunt protuberances from the edge of a thunderstorm echo. This technique of identifying hail shafts assumes that large hail is characteristically a fringe phenomenon of a thunderstorm. This assumption may not be valid. Hence, these indicators appear to be exceedingly qualitative and unreliable, and more objective and conclusive characteristics are required to identify hailstorms.

A study of echo motions in severe storms in the midwestern United States by Stout and Hiser (1955) showed a high degree of association of echo convergence and vortex motion with severe winds and hail. This motion is revealed by echo mergers, curling motion, an appendage, echo rotation, or line rotation. Such echo characteristics may serve as an alerting signal for a forecaster but have not been tested sufficiently to provide the criteria for hailstorm identification.

6.2 Reflectivity and height:

Vertical profiles of radar reflectivities from hailstorms have been reported by Donaldson (1961) and Geotis (1963) in New England, Atlas and Ludlam (1961) in England, Douglas (1963) and Chisholm (1968) in Alberta, Canada, Changnon and Stout (1964) in Illinois and Dennis et al. (1970) in South Dakota. These observations were taken either with 3-cm radars, having a serious problem of attenuation or with 10-cm radars, where attenuation is negligible.

6.2.1 New England hailstorms:

Extensive measurements of radar reflectivities at various heights in the cores of New England thunderstorms have been reported by Donaldson (1961). These were made with a 3.2 cm wavelength CPS-9 radar located at Milton, Massachusetts. The surface weather conditions were furnished by an extensive network of over 500 volunteer observers, located in southern New England. The core profiles, numbering 233, were correlated with surface observations, the condition being that heavy rain or hail must be falling during the time of profile measurement or within the following fifteen minutes.

The median profiles of core Z_e (equivalent Z) in 1957-1958 New England thunderstorms are illustrated in Fig. 6-1. The radar reflectivity factor Z is a measure of the scattering power of a group of particles, and it is defined as the summation, over a unit volume, of the sixth power of particle diameters. The definition of Z_e of a radar echo is simply the Z of small (Rayleigh-approximation) spherical water drops that would have the same total backscattering cross section as the measured echo. The profiles are arranged in category of most severe weather. They show that their $Z_{e_{max}}$ and its height both increase with the severity of the thunderstorm. The largest differences were found at atltitudes of about 7.5 km and higher. Hence Donaldson suggested that <u>most severe thunderstorms producing hail or tornadoes may be reliably identified by radar reflectivities in the upper half of the storm</u>. A detailed analysis of the frequency distribution of core Z_e at various altitudes confirms this even better than the medians. The different types are scarcely distinct at 1.5 km altitude. However, the Z_e curve for hail exceeds the curve for rain by an order of magnitude or more at 6.0 and 8.0 km altitudes, except for the most intense part of both distributions. This may be due to incomplete reporting of hail in even a rather dense observing network.

These observations are well confirmed in various locations by other investigators, such as Ludlam (1959), Browning and Ludlam (1960), Wilk (1960) and Inman (1960). <u>Their observations</u>

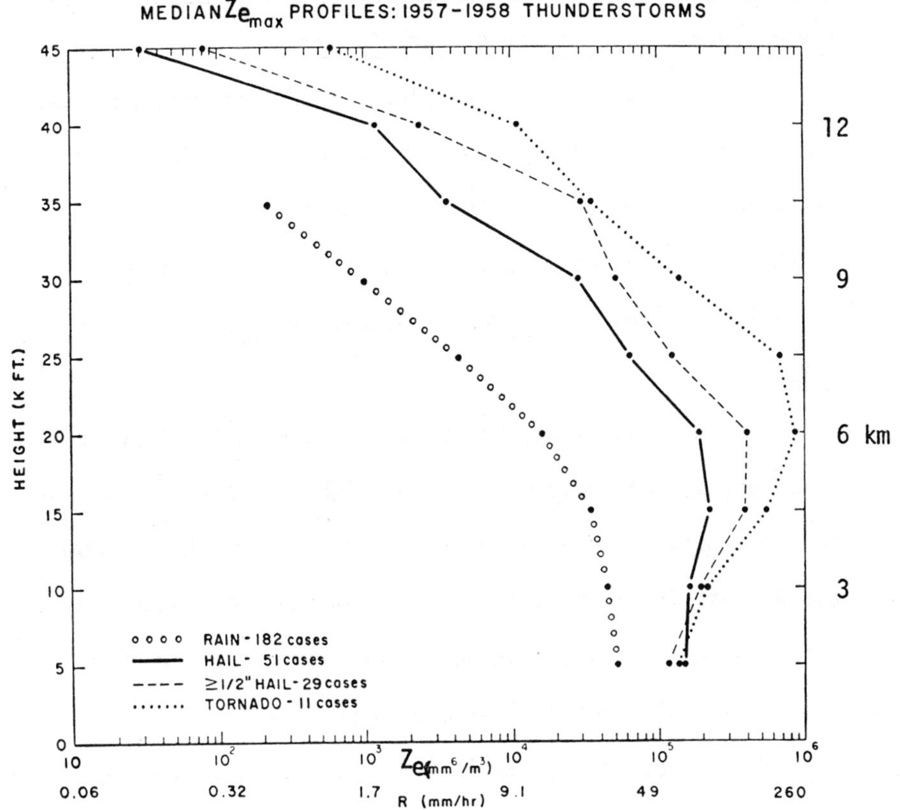

FIG. 6-1 Median profiles of core Z_e in 1957-58 New England thunderstorms, arranged in category of most severe weather. The 51 cases of hail include the 29 cases of large hail [diameter of 1.2 cm or larger] which are plotted separately. Also, the 11 tornado profiles are taken from the all-inclusive rain and hail categories. (from Donaldson, 1961)

show intense core echoes for hailstorms with a reflectivity maximum generally between 4.5 and 9.0 km.

The extremely high values of reflectivity observed aloft are interpreted by Donaldson as an indication of storage of large hailstones in concentrations of 10 g m^{-3} or more. Further, these values provide evidence for the storage aloft of hailstones near the level of updraft maximum in a convection cell.

Douglas (1963) has plotted the probability of hail as a function of maximum echo top height as shown in Fig. 6-2. The probability increases with the increase in the maximum echo top height, in Alberta, Texas, and New England. When echo tops reach heights in excess of about 9 km above terrain, the probability of large hail becomes significant.

In contrast, Geotis (1963) reported that echo tops of hailstorms which he observed in New England do not show any correlation with hail size or storm duration. Some storms which were only 6 km in height produced hail as large as 2.5 cm in diameter and track lengths of 16 km or more. Moreover, tall storms do not necessarily produce large hail, as seen from Fig. 6-3. The 10-cm radar reflectivity of thunderstorms, on the other hand, is a good indicator of hail and a rough indicator of its size, as shown in Fig. 6-4. No significant characteristics were noted for hailstorms except their significantly high reflectivity. In New England thunderstorms, if the radar reflectivity (Z_e) exceeds

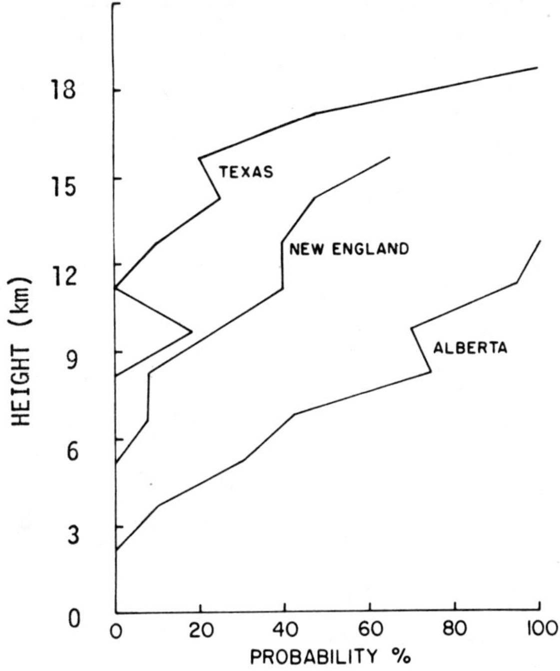

FIG. 6- 2 The probability of hail as a function of the maximum echo top height. [Texas data from Inman, private communication.] (from Douglas, 1963)

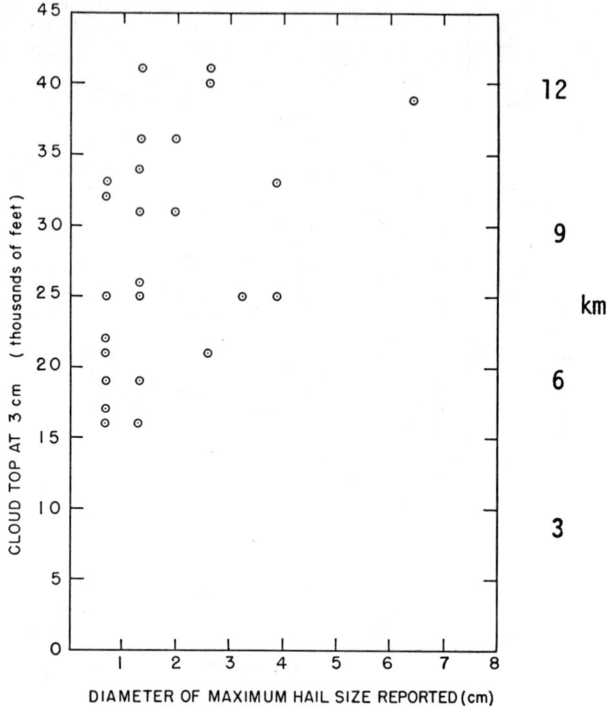

FIG. 6- 3 Maximum reported hail diameters from track-producing storms of 1961 as a function of storm height [measured at 3 cm]. (from Geotis, 1963)

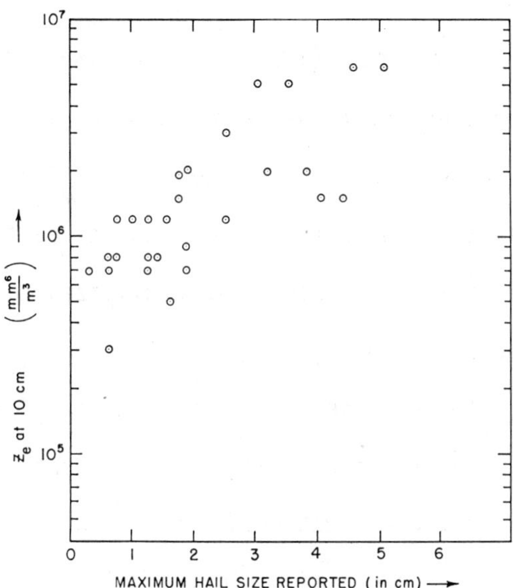

Fig. 6-4 Maximum measured Z_e at 10 cm versus maximum reported hail diameter for the track-producing storms of 1961. (from Geotis, 1963)

$10^{5.5}$ mm^{6}m^{-3}, the hail very likely will reach the ground. If this value is exceeded for several minutes, Geotis concludes that hail at the ground level is certain.

6.2.2 Wokingham Storm:

Atlas and Ludlam (1961) and Browning and Ludlam (1960, 1962) made an intensive radar study of an individual hailstorm in and near the town of Wokingham. This storm traveled across England from the south to the east coast and produced large hail. Five radars were utilized to survey the storm with wavelengths of 3.3, 4.7 and 10 cm.

The intense phase of the storm lasted for more than 3/4 hour and golf-ball size hailstones fell continuously. The features of the radar echo structure are shown on a RHI section through the approaching storm in Fig. 6-5. A height of more than 14 km was reached by the radar echo top, and maximum echo intensity of 71 dB was attained at 8 km (Fig. 6-5). Browning and Ludlam have particularly emphasized that the echo moved uniformly throughout this phase, maintaining the different features shown in the diagram, such as the 'wall,' 'anvil echo,' 'forward overhang' and the 'echo-free vault.' Their characteristics are quoted below as described by Ludlam (1963).

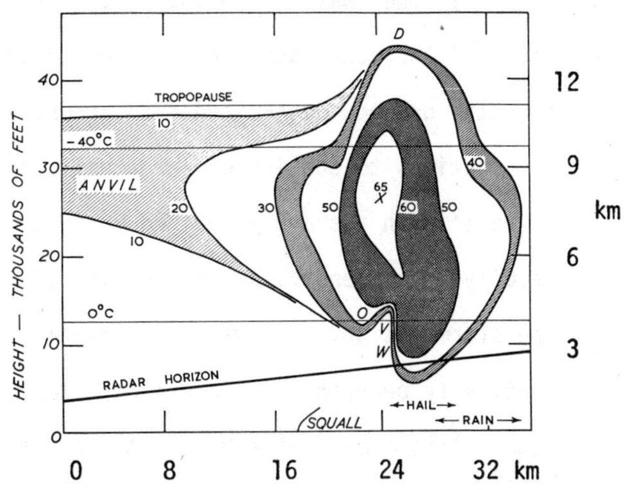

FIG. 6-5 Features of radar echo structure in the Wokingham storm during its intense phase, shown on a range-height section through the approaching storm. Isopleths are drawn of 4.7 cm reflectivity expressed as 10 log Z_e ($mm^6\ m^{-3}$). X: position of maximum radar echo intensity; D: highest echo top, almost vertically above W, the echo 'wall,' associated with large hail. O: the 'forward overhang,' separated from the wall by V, the 'echo-free vault.' (from Ludlam, 1963)

"a) The 'wall.' In the central right-hand parts of the echo mass the leading precipitation near the ground, determined from ground observations to consist of large hail, formed a 'wall' with a sharply defined upright front face lying perpendicular to the direction of the storm motion. The wall was practically vertically beneath the highest echo top, and rather less than 1 mile behind the position of

"b) the most intense echo at 3.3 and 4.7 cm, centered at a height of about 25,000 ft. Atlas and Ludlam (1961) showed that at this position the magnitude and the variation with wavelength of echo intensity point to the presence of hailstones with an almost uniform diameter of about 5 cm and with dry surfaces (stones of this size reached the ground over areas about 1 km across).

"c) The 'anvil' echo. Diffuse echo extended forward of the storm, evidently corresponding to an anvil cloud. Its top was at about 35,000 ft and its base descended gradually from about 25,000 ft 25 miles ahead of the storm to form a "d) 'forward overhang' of echo down to about 12,000 ft 1-3 miles ahead of the wall and along its length. Between the forward overhang and the wall the base of the echo rose again to 15,000 ft or even more, enclosing

"e) the 'echo-free vault,' which Browning and Ludlam (1962) regard as marking the zone in which a strong updraft enters the front of the storm and prevents the descent of the particles composing the base of the forward overhang. From the low echo intensity there it appeared that they consisted of small (millimetric) hail whose fall-speeds did not exceed about 10 m sec^{-1}."

6.2.3 Illinois hailstorms:

A large number of observations of hail, damaging winds and rain taken in central Illinois during 1958 and 1960 were analyzed by Wilk (1960, 1961) in conjunction with radar reflectivity observations taken with the CPS-9 radar, which was an experimental model without the RHI capability. The antenna was tilted through a series of 5 tilt angles for making reflectivity measurements. The highest Z_e value frequently observed for hailstorms was from $1x10^5$ to $9x10^5$ mm^6m^{-3}, and was located between the 4.5 km and 6.0 km heights.

During the afternoon of August 8, 1963 near Weldon, Illinois, a storm produced hail over an area of approximately 44 km^2. An exhaustive study of it has been reported by Changnon and Stout (1964). The storm was tracked by two of the Water Survey's radar sets, one CPS-9 and the other TPS-10. The hailstorm cell was detected from 32 to 48 km away from the radar sets. The scope presentations, one PPI and the other RHI, of these radars were photographed furnishing a three-dimensional portrayal of the storm cell revealing the precipitation development in it. The first echo of the hail cell observed at 1544 CST, extended in the vertical from 9 km to about 13 km as seen in Fig. 6-6. Such a high level echo development is rather uncommon for thunderstorms. Browning et al. (1963), however, reported similar heights for first echoes of hail-producing storms in Oklahoma. Figure 6-6 shows the three-dimensional models of the RHI-radar echoes and the development of the hailstorm. By 1550 CST at least three turrets extended above the main echo. The important characteristics of the hailstorm echo development were: the 'first' echo development at unusually high levels indicating strong updrafts and echo due to ice particles, the diverse cellular structure and the presence of shear at the 3.5 km level.

Towery et al. (1970) have reported limited results

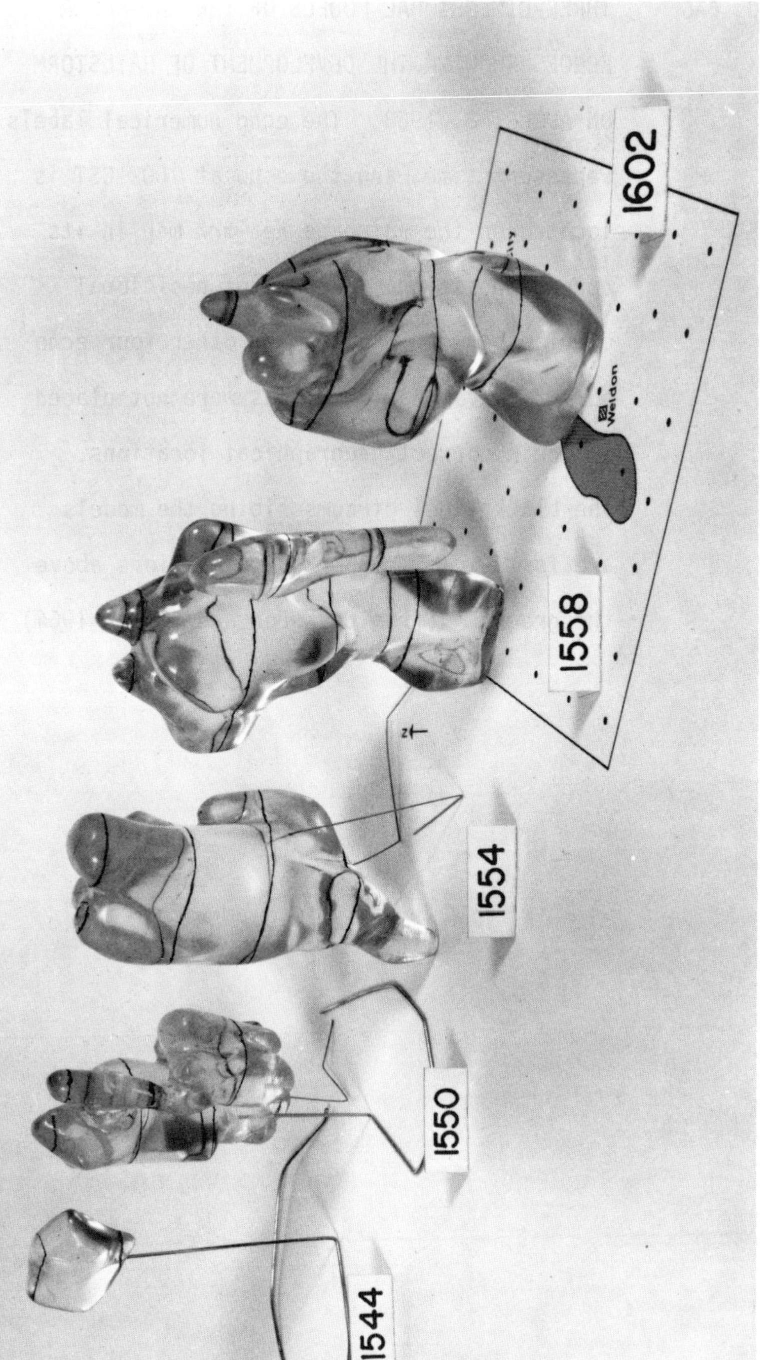

FIG. 6-6 (Figure legend on next page)

FIG. 6-6 THREE-DIMENSIONAL MODELS OF THE RHI-RADAR ECHOES SHOWING THE DEVELOPMENT OF HAILSTORM ON AUGUST 8, 1963. The echo numerical labels represent time, and the echo at 1602 CST is located on the raingage network map in its correct position. Because of positional overlap between echoes, the other four echo portrayals at earlier times were not placed in their correct geographical locations. The black lines circumscribing the models are for 3, 6, 9, and 12 km elevations above the ground. (from Changnon and Stout, 1964)

of RHI data for two hail days in 1969. A rapid increase was noted in radar reflectivities (10^3 through 10^5 mm^6m^{-3}) in the five minute period before hail reached the ground. The echoes not associated with hail also showed similar characteristics. However, their volumes (with $Z_e \geqq 10^5$ mm^6m^{-3}) were much smaller and short-lived, as compared to those of hail echoes. A very interesting and important conclusion was drawn that a critical volume of high liquid water content must precede the hail production by several minutes.

6.2.4 Hailstorms of Western South Dakota:

The results of an extensive observational program to gather data on the hailstorms of Western South Dakota have been reported by Dennis et al. (1970). Data were obtained from five aircraft, a network of hail indicators and raingages, a mobile observer, time lapse cameras and four radar sets. In addition, quantitative radar data were obtained with a 3.2 cm set. A computer was used to generate synthetic PPI, RHI and HARPI (height-azimuth-range position indicator) scans. The values of 10 log Z_e, which were not corrected for attenuation, were presented at the points of a grid. The observations were made for 22 storms in which hail was considered likely.

The multicellular nature of the storms was invariably observed. On the radar screen, cells lasted as separate entities for $\sim$ 20 - 30 minutes. However, a few cells of steady state characteristics lasted much longer, almost an hour. The inflow at the trailing edge of the hailstorms was detected in a number of cases. Most active cells were almost vertical, and these indicated pronounced slope due to wind shear only in their dissipating states. The distributions of maximum echo height, maximum equivalent reflectivity factor Z_e, and height of maximum Z_e are shown in Fig. 6-7. The median observed value of maximum Z_e of 48 dB for the hailers is low in comparison with values reported by earlier observers. It may be due to the value not being corrected for attenuation.

Vertical profiles of Z_e at various times in cores of cells of a large storm are shown in Fig. 6-8. The first echoes appear in new cells at a high level around 10 km. The peak begins to decrease but not before it has reached a value which exceeds 50 dB. This suggests a considerable amount of precipitation temporarily supported at upper levels by strong updraft. A typical value of the descent rate was 10 m sec^{-1}. The peak took about 15 to 25 minutes to arrive at the ground.

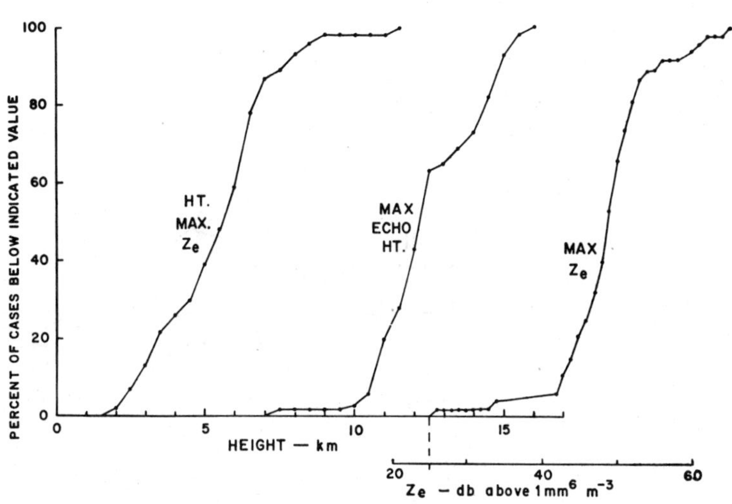

Fig. 6-7 Distributions of maximum echo height, maximum equivalent reflectivity factor Z_e, and height of maximum Z_e for 15-min time blocks with hail in 1968. (from Dennis et al., 1970)

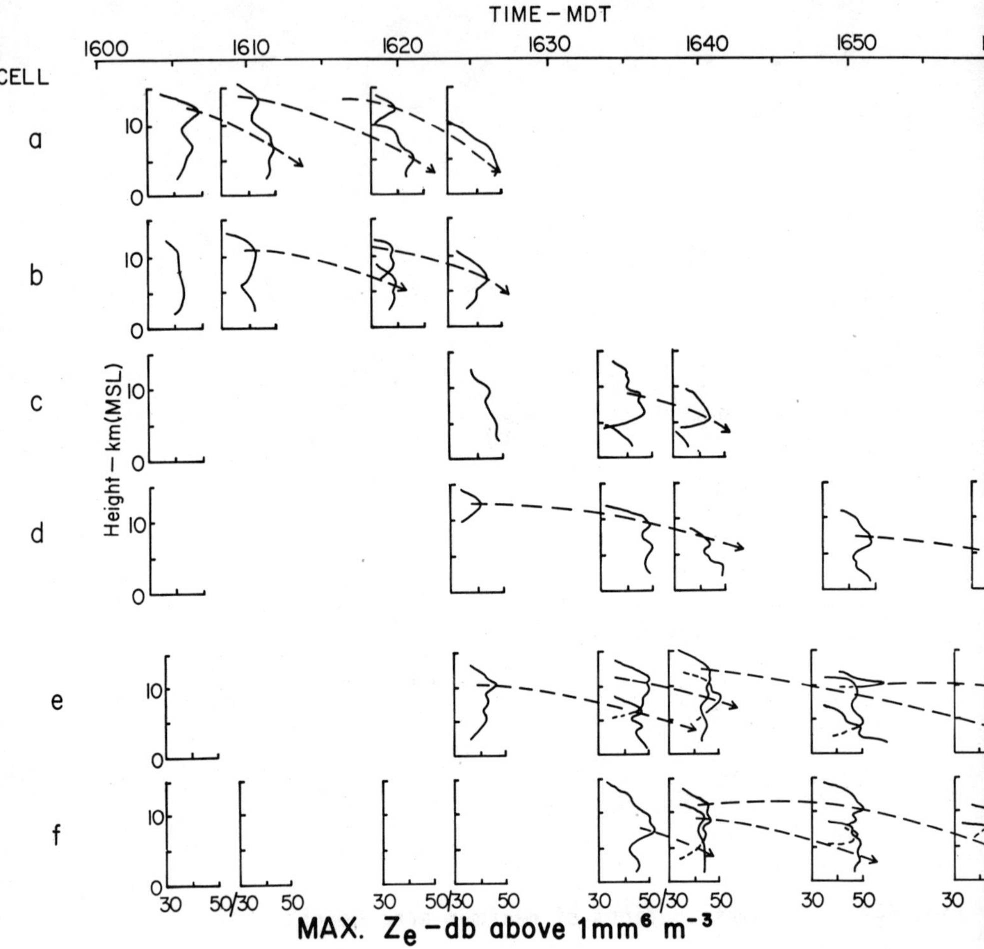

FIG. 6-8 Vertical profiles of Z_e at various times in cores of cells. (from Dennis et al., 1970)

6.2.5 Alberta hailstorms:

The details of the hail reporting network in Alberta, Canada, have been discussed in Chapter 3, section 3.1.3. The patterns of the radar echoes of hailstorms studied by this group (Hitschfeld, 1971) with the help of a reporting network and three radars, namely, 3.2 cm, broad-beam 10 cm and high resolution 10 cm, are discussed in this section.

Their original 3.2 cm radar had good vertical resolution and provided precise information on the height of the echoes. Figure 6-9 shows good correlation between hail incidence and maximum echo height, agreeing well with Donaldson's (1958) findings. However, their storm echoes suffered from attenuation, and therefore, the measurements of reflectivity factor Z were not useable quantitatively. In spite of this serious limitation, correlations of hail incidence with height, temperature, top of the first echo, and liquid water contents of echo tops were attempted (Figs. 6-9 and 6-10).

The radar facility was considerably improved in 1967, when a 10 cm radar was equipped with a big parabolic antenna, resulting in a conical beam pattern of 1.15 degrees. This

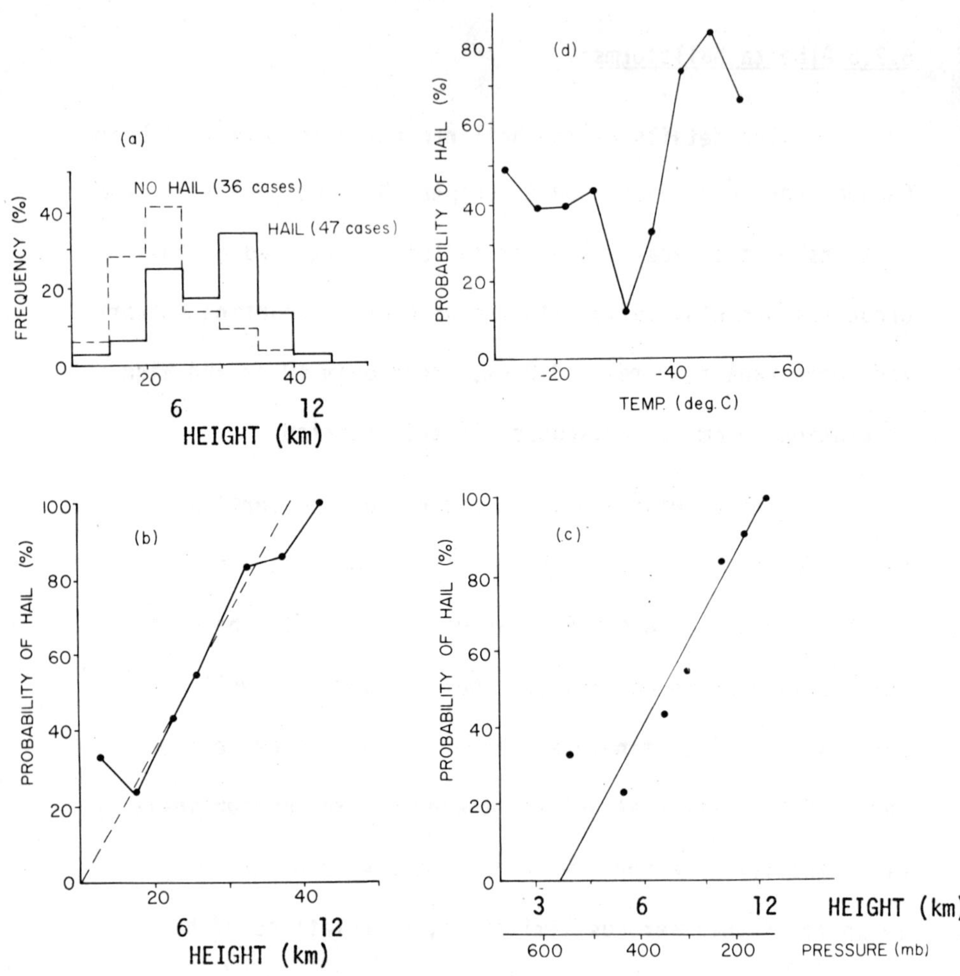

FIG. 6- 9 (a) Distribution with height of 47 storms which produced hail, and of 36 storms which did not. (b) Probability of hail fall [number of hailstorms of given height divided by total number of storms of that height], plotted against maximum echo height above surface. (c) Probability of hail plotted against logarithmic scales of maximum echo height. (d) Probability of hail plotted against temperature prevailing at echo top. (from Douglas and Hitschfeld, 1958)

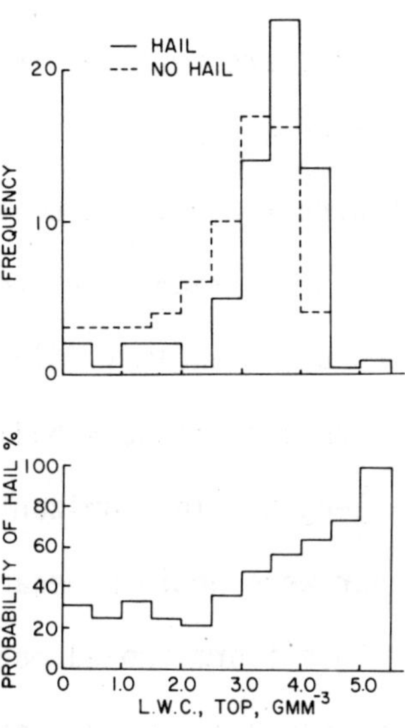

FIG. 6- 10 Frequency distributions of liquid water contents of echo tops for both hail producing and non-producing echoes. The bottom diagram shows the [smoothed] probability of obtaining hail as function of the water content in the echo top. (from Douglas, Marshall and Barklie, 1962)

antenna scans the atmosphere up to 20 degrees elevation and takes 3 minutes to complete the cycle. A simplified grey scale system was introduced to show five intensity levels at 10 dbZ intervals. Two major types of storms are reported from these observations: the long-lasting quasi-steady storm, and smaller, short-lived sporadic cells which were found to occur sometimes in groups. The storm of the first type which occurred in Alberta on 29 June 1967 has been reported by Chisholm (1968). The 10 cm radar echo persisted for five hours and produced hail of golf-ball size over a path 96 km in length. The horizontal and vertical resolution of the radar were optimum because the storm was less than 48 km from it. This storm showed many characteristics similar to those of steady state storms observed by Browning and Ludlam (1962), Donaldson (1962) and Browning (1965). The interesting configuration and reflectivity structure are displayed in Fig. 6-11, which remained constant for a long time, especially after the 'vault' was evident. The 'vault' is an echo-free region which was visible for more than 2 hours, and had a width of about 8 km in diameter at its base. In Fig. 6-11, the maximum Z_e values are labeled as positive and the minimum values as negative. Figure 6-12 is a vertical cross section taken along line A-B. It also shows the vault area and the overhang. A cross section taken along line C-D, which is perpendicular to the line A-B, has similar characteristics.

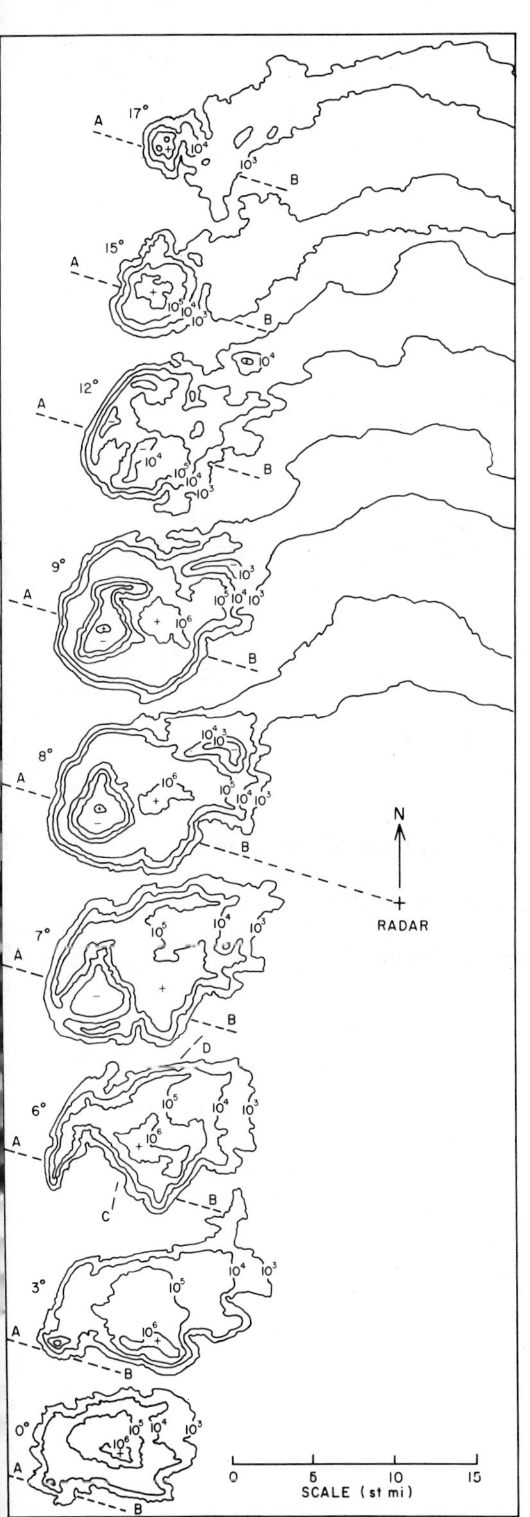

FIG. 6-11 PPI view of storm 1356-1359 MST. PPI sections are shown for elevations 0-17°. Contours are $Z_e(mm^6m^{-3})$. (from Chisholm, 1968)

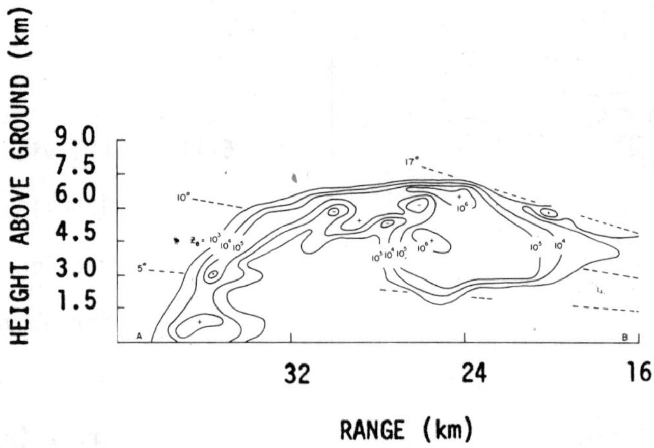

Fig. 6-12 Vertical cross section along line A-B. Contours are $Z_e (mm^6 m^{-3})$. (from Chisholm, 1968)

Figure 6-13 shows a series of core reflectivity profiles. The average values over 15 minute periods are plotted between 1230 and 1400 MST. The interesting feature common to these profiles is that they are all essentially constant with height up to 4.5 km. The values decrease markedly at upper levels. This type of reflectivity variation was reported earlier by Geotis (1963) and Proctor and Geotis (1963) in hailstorms in New England. With the exception of one profile, all others show values between 10^6 and 10^7 mm^6m^{-3} up to the 4.5 km height. Figure 6-14 displays the plot of observed rain and hailstone sizes against maximum reflectivity values (Chisholm, 1968). The rain reports indicate Z_e values between $8x10^3$ and 10^5 mm^6m^{-3}. The vast majority of hail reports, especially for smaller size hailstones, are within the $Z_e = 10^5$ mm^6m^{-3} contour. The walnut or golf-ball size hailstones are close to a $Z_e = 10^6$ mm^6m^{-3} contour. Thus there is a definite increase in the Z_e value as the size of hail increases from shot to walnut. A similar finding has been reported by Geotis (1963).

6.3 Significance of the reflectivity maximum aloft:

As discussed in the previous section, a number of investigators have reported radar reflectivity profiles in hailstorms, notably Browning and Ludlam (1960) and Atlas and Ludlam (1961) in England, Donaldson (1961) and Geotis (1963) in New England, Wilk (1960) in Illinois, Dennis et al. (1970) in South Dakota, and Douglas and Hitschfeld (1958) and Chisholm

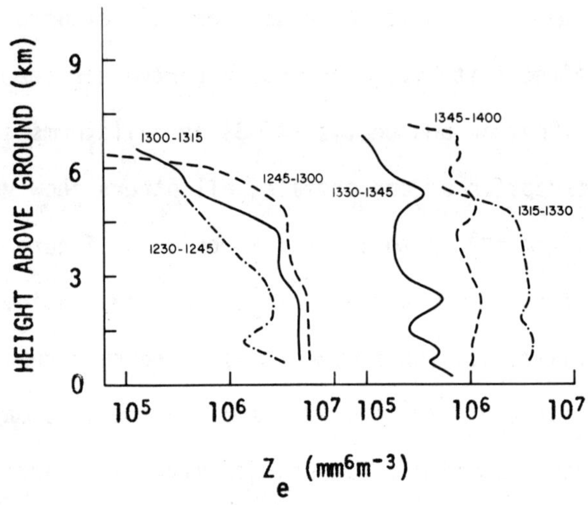

FIG. 6-13 Core reflectivity profiles. Average reflectivity over 15 min periods is given for 1230-1400 MST. (from Chisholm, 1968)

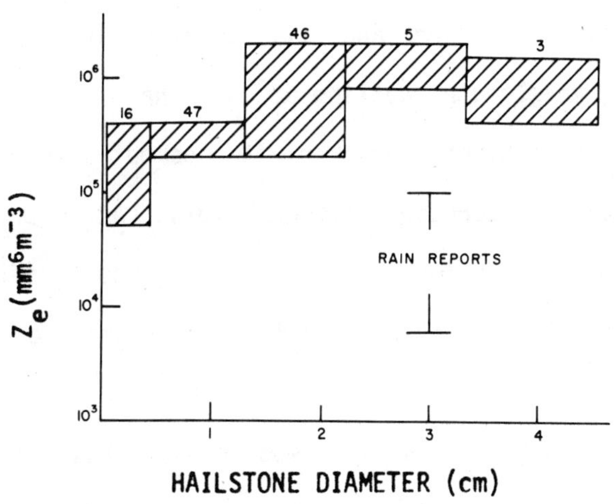

FIG. 6-14 Observed hailstone sizes and maximum reflectivity values. Numbers above hatched areas indicate the total number of hail reports used. There were 79 rain reports. (from Chisholm, 1968)

(1968) in Alberta, Canada. In summary, we can say that these observations show that hailstorms have intense core echoes at middle altitudes, with a reflectivity maximum generally between 4.5 km and 9 km. The reflectivity maxium aloft is probably due to the following three factors (Atlas, 1963): (i) attenuation increasing towards the base; (ii) reduction in radar cross section of large hail due to melting; and (iii) greater particle concentrat aloft. At 3 cm wavelength, wetting of large hail as well as attenuation is significant. However, it does not account for all the reflectivity decrease at lower levels. At 10 cm, both wetting and attenuation are negligible. Hence, reflectivity maximum aloft at 10 cm must indicate the accumulation of hydrometeo at upper levels.

Donaldson (1961) as discussed earlier, has interpreted the extremely high values of reflectivity in the upper regions of a few thunderstorms as indicators of large hailstones. His study also provides evidence for the accumulation of hydrometeors near and above the altitude of updraft maximum. The accumulation occurs in the small core region of the hail cell and in a limited altitude range of about 6 km. A gradual release of particles takes place when the cell is in the dissipating stage and a new cell then takes over.

When an intense updraft develops rapidly, the first echo appears at a very high elevation and then descends slowly towards the base of the cloud as indicated by the studies of

Arnold (1961), Changnon and Stout (1964) and Dennis et al. (1970). Such echo development indicates major growth of hydrometeors while they fall slowly towards the updraft maximum. Thus, major growth takes place in the upper part of the updraft while the hydrometeors are continuously suspended due to the characteristic profile of the updraft. This topic is discussed in more detail in Chapter 8.

6.4 Multiwavelength techniques for hail identification and sizing.

Atlas and Ludlam (1961) have made computations of the equivalent radar reflectivity (Z_e) of hailstones of a single size. In addition to size, the value of Z_e depends on the radar wavelength, and the dryness or wetness of the particles. Figure 6-15 shows the values of 10 log Z_e for a concentration of 1 g m^{-3} of spherical dry hailstones of indicated sizes, at three wavelengths, 3.3 cm, 4.7 cm and 10 cm. Figure 6-16 is similar to Fig. 6-15 except that the hailstones are electromagnetically wet. An exponential curve fitted to the maximum concentrations of largest hailstones is shown in Chapter 2 (Fig. 2-23). Assuming hail size spectra of varying slope, the total reflectivity and the equivalent reflectivity factor Z_e calculated by Atlas and Ludlam (1960) are shown in Fig. 6-17. The variation of Z_e with wavelength is significant, especially for wet hail.

Thirty-three samples of hailstones were collected in Alberta, (Douglas and Hitschfeld, 1961) from a wide area on 18

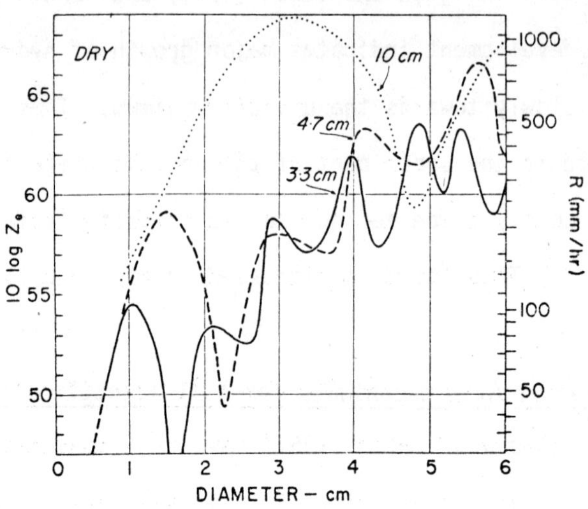

FIG. 6-15 10 log Z_e (Z_e mm^6m^{-3}) for a concentration of 1 g m^{-3} of spherical hail of the indicated sizes, assuming the hail is electromagnetically dry at the wavelengths shown. Equivalent rainfall rate based on $Z_e = 486R^{1.37}$. (from Atlas and Ludlam, 1961)

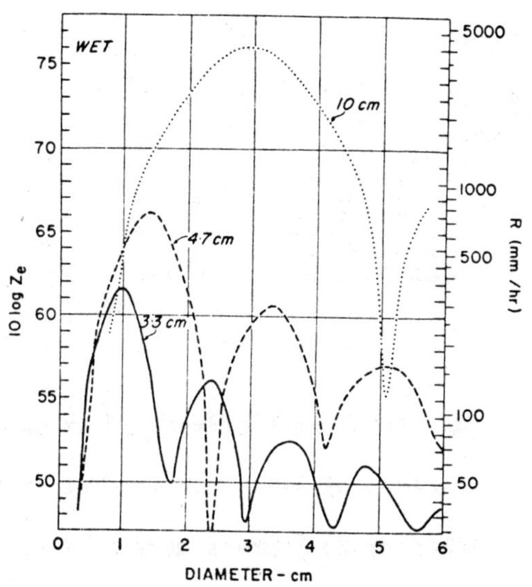

FIG. 6-16 Same as Fig. 6-15 but the hail is electromagnetically wet. (from Atlas, 1964)

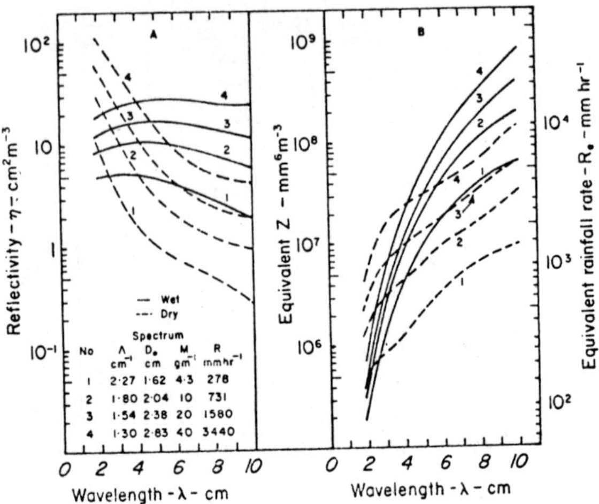

FIG. 6-17 (A) Total reflectivity of an exponential size spectrum of hail such as that in Fig. 2-22 (Chapter 2) [curves marked 1] as a function of wavelength. Solid curve for wet hail; dashed curve for dry hail. (B) Equivalent reflectivity factor Z_e corresponding to curves in A. Numbers on curves indicate exponential spectra of varying slope Λ as shown in the legend. (from Atlas and Ludlam, 1961)

different days. The samples represent hailfall rates (R_{melted}) over a range 0.02 to 40 mm hr^{-1} and ice concentrations (M) aloft in the range 0.01 to 6 g m^{-3}. Using the equations suggested by Atlas and Ludlam (1960), the equivalent reflectivity Z_e was computed by Douglas and Hitschfeld (1961) for each sample. The relevant relationships among Z_e, R and M for hail, rain and snow are summarized in Table 6-1. It is interesting to note that the reflectivities at 3.3 cm wavelength are greater for wet than for dry hail by a factor ranging from two to six. Only for a few samples of large hailstones, the reverse is true. The reflectivities are higher for wet stones than for dry ones in all cases for λ equal to 10 cm. Thus a fairly sensitive dependence of Z_e exists on parameters such as hailstone size, concentration of hail, and wavelength in the microwave band. However, the effect of surface wetness, roughness, shape and orientation so complicate the reflectivity-wavelength dependence as to render measurements on hail size, meaningless.

Similar studies in the Soviet Union have been reported by Sulakvelidze (1966a). The results are summarized in Fig. 6-18. The values of radar reflectivities are plotted against the diameter of hailstones for two wavelengths (3.2 and 10 cm) and for different concentrations. Sulakvelidze concludes that if the contration of hailstones in a cloud is known, their size can be definitely determined from the reflectivity value at 10 cm. But since

TABLE 6-1

Relationships among Z_e, R and M for hail, rain and snow.

(from Douglas and Hitschfeld, 1961)

(Z_e mm^{6}m^{-3}, R mm hr^{-1}, M mgm m^{-3})			
	Z_e and R	Z_e and M	M and R
Hail, wet, λ=3.3 cm	Z_e=3.5x10^{4}R$^{0.88}$	Z_e=8.5x10^{2}M$^{0.85}$	M =105 R$^{0.9}$
" dry "	=8x10^{3}R$^{1.05}$	=1.1x10^{2}M$^{1.05}$	
" wet λ= 10 cm	=6.2x10^{4}R$^{1.37}$	=3.3x10^{2}M$^{1.25}$	
" dry "	=1.5x10^{4}R$^{1.21}$	=1.6x10^{2}M$^{1.10}$	
Rain (ref.3)	Z_e=2 x 10^2 R$^{1.6}$	Z_e=8.3x10^{-2}M$^{1.82}$	M = 72 R$^{0.8}$
Aggreg. Snow "	Z_e=2 x 10^{3}R$^{2.0}$	Z_e=9.6x10^{-3}M$^{2.2}$	M =250 R$^{0.9}$

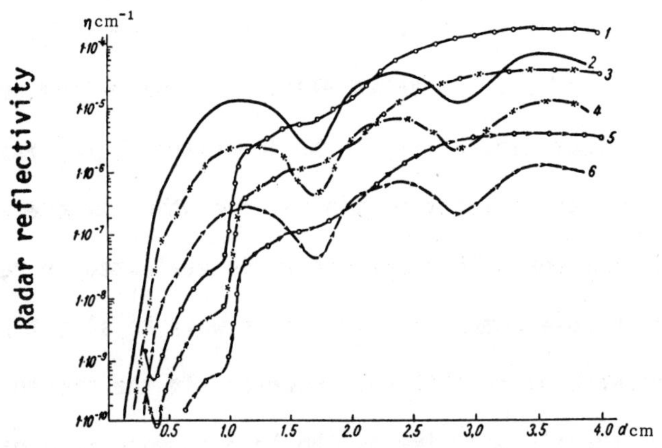

Number of curve	N, cm^{-3}	λ, cm	Number of curve	N, cm^{-3}	λ, cm
1	$1 \cdot 10^{-5}$	10	4	$1 \cdot 10^{-6}$	3.2
2	$1 \cdot 10^{-5}$	3.2	5	$1 \cdot 10^{-6}$	10
3	$1 \cdot 10^{-6}$	10	6	$1 \cdot 10^{-7}$	3.2

FIG. 6-18 Graph of radar reflectivity against diameter for wavelengths λ of 3.2 and 10.0 cm and concentration N of 0.1 and 10 m^{-3}. (from Sulakvelidze, 1966a)

hailstone concentration in a cloud is unknown, determination of the size, even of monodisperse hailstones, is impossible.

Eccles and Atlas (1973) recently proposed a new method of hail detection, which functions on the principle of scattering properties of hail which are distinctly different from raindrops at the 10 cm and 3 cm radar wavelengths. The method consists of taking the range derivative of the logarithm P_{10}/P_3 ratio of received powers and thereby eliminates the abiguity due to 3 cm signal attenuation. This derivative is evaluated in real-time by a special purpose computer. It will always be slightly positive in the presence of rain alone. However, in the presence of hail it becomes sharply positive at the leading edge of a hail shaft and sharply negative at the trailing edge, which is used as an indicator of hail.

The preliminary test of the dual wavelength hail detector was very recently carried out. There are, however, problems in accurate evaluation of the above mentioned derivative (Carbone, 1971, 1972; Srivastava and Carbone, 1971). For example, a measurement of the derivative may deviate from its average value because of random fluctuations in signal powers, and its becoming negative may give a false indication of hail or a 'false alarm.' Other factors influencing the magnitude of the derivative may be the effects of finite pulse length, finite range increments between successive pulse volumes, background rainfall rates, hail concentration gradients, hail wetness, the plan view dimensions of the hail shaft, finite beam width, and the presence of non-hail non-Rayleigh scatterers such as large raindrops and melting snow and still unrecognized features of the signals and their statistics. It is not obvious how such factors will combine to affect the Eccles-Atlas detection scheme. In view of the aforementioned difficulties, further detailed analysis and field experimentation are necessary to determine the usefulness, sensitivity and false alarm rate of such hail detection.

6.5 Other techniques:

6.5.1 Polarization:

Polarization techniques for hail detection (Atlas and Wexler, 1963) depend upon the fact that nonspherical hailstones generally respond differently to vertical and horizontal polarization. Thus, the variations in echo intensity associated with polarization switching from the vertical to the horizontal should theoretically indicate the presence of oriented large hail rather than spherical raindrops. It is known, however, that large suspended raindrops are not spherical and frozen particles experience considerable tumbling during their growth. Switching from linear to circular polarization may provide another means of identifying large nonspherical particles. This technique, however, has been tried in air-borne 5.5 cm weather radar without success by Harrison and Post (1954). But recent studies by Barge (1970,

1972) indicate some evidence that deformed hail may be detected by its depolarization signature.

6.5.2 Use of pulsed-Doppler radar:

Pulsed-Doppler radar shows great potential for the quantitative determination of particle size distributions and the measurement of vertical motions in thunderstorms and the large scale mapping of wind field (Battan and Theiss, 1966). Very recently, Battan and Theiss (1972) have taken Doppler radar observations during hailfall. Their analysis of the data indicates that the variance of the spectrum of the vertical Doppler velocity is a fairly good indicator of maximum hailstone sizes (Donaldson and Wexler, 1969; Boston and Rogers, 1969). However, there exist a number of uncertainties in this procedure, such as the estimation of updraft velocity, particle shape, composition and fallspeed. Thus this method, though not an operational technique yet, points the way to interesting possibilities.

CHAPTER 7

MODELS OF HAILSTONE GROWTH

Different environmental conditions are conducive to the formation of a variety of thunderstorms, ranging from short-lived convective cells which form in a random fashion, to the steady state and organized severe storms which have a longer life. The earlier storm models, which are well described by Ludlam (1963), express the different ideas of the air flow and precipitation distributions in a mature cumulonimbus cloud. The important features suggested in these models are: (i) vigorously developing turrets, extending high into the upper atmospheric layers; (ii) presence of anvil cloud, indicating the effect of wind shear and upper level divergence; (iii) ascent of air in front of the storm; and (iv) development of downdrafts due to organized circulation, falling of precipitation and partial chilling due to evaporation and melting of ice particles. These features are shown in Fig. 7-1. For more details of these models, the reader may refer to the original papers by Möller (1884), Davis (1894), Wegener (1911), Brooks (1922), Simpson (1925), Letzmann (1930), Suckstorff (1939) and Findeisen (1940).

Technological advances in radar, radiosondes, balloons, instrumented aircraft and high speed computers have helped in securing improved analysis of the temperature, moisture, and pressure distributions, the wind patterns, and the microstructure

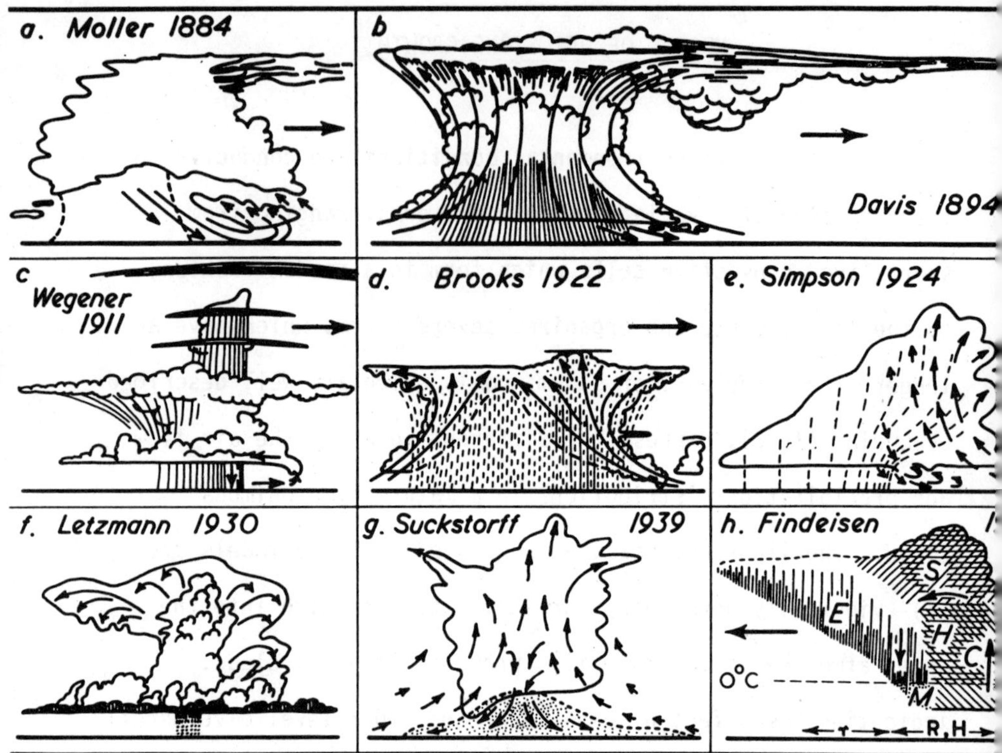

FIG. 7-1 Cumulonimbus models. The direction of motion of the storm is given by the heavy arrow. a) In Möller's model the downdraft air ascends at the storm front. b) This model emphasizes the great extension of the anvil and its mamma. c) The anvil is less marked a even partly detached: precipitation near the ground is most intense at the storm front. d) The central region contains precipitation <u>only</u> (dashes), beneath inclined cloudy updrafts. The dome of the anvil is surmounted by pileus; hail (vertical lines) occurs below it, near the storm front and its arch cloud.

(continued next pa

FIG. 7-1 continued

e) There is no realism in the shape, but the flow of air and precipitation is marked clearly and perceptively. f) Improved realism of shape, but little intrusion of dynamical ideas. g) Comparative symmetry; a model from the tropics. h) Microphysical processes introduced into the asymmetrical cumulonimbus; oblique hatching: condensation (C) or sublimation (S); horizontal hatching: hail growth (H); vertical hatching: ice particle evaporation (E) or melting (M). Feeble rain (r) ahead of the region of heavier rain and hail nearer the updraft. (from Ludlam, 1963)

of the storms. Major storm models proposed in a comparatively recent period are discussed by Atlas et al. (1963), Newton (1963), Ludlam (1958, 1963), Aufm Kampe and Weickmann (1957) and Weickmann (1957). The features of these cumulonimbus models will not be discussed here except those which apply specifically to the development of hail in such storms.

7.1 Thunderstorm project:

A field study of significant and far-reaching importance was carried out by Byers and Braham (1949) in Florida and Ohio, namely the well-known Thunderstorm Project. It supplied for the first time many details of the three dimensional structure of large cumulus clouds and relatively small air-mass type thunderstor Robust aircraft capable of traversing thunderstorms, radar and other meteorological tools were extensively utilized to obtain the data. The model which resulted from the new quantitative information gathered during this project has great merit, as it emphasized the idea of storm evolution. Among the most important findings are those presented in subsequent paragraphs.

7.1.1 Random and squall-line thunderstorms:

Figure 7-2 shows a schematic representation of the radar echoes on a day with randomly-distributed thunderstorms compared with the echoes due to a squall line. The black areas (storm echoes) are regions of net upward motion and the white

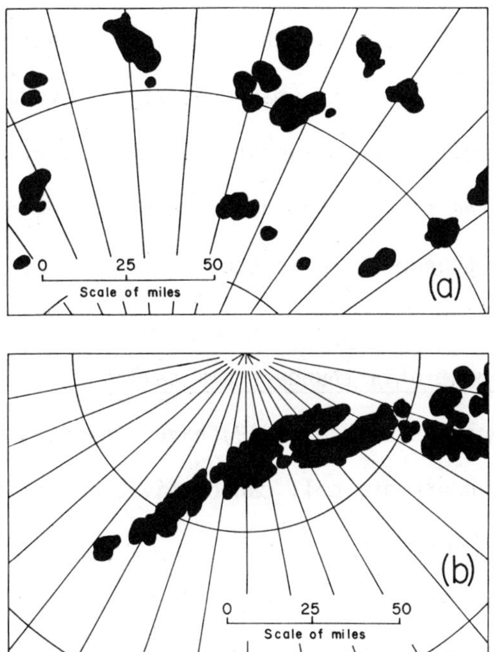

FIG. 7-2 Radar echoes in cases with random (a) and squall-line (b) thunderstorms. (from Byers and Braham, 1949)

areas are the regions of net downward motion (often cloud-free). The downflow may not, however, extend throughout the entire region between upflow areas. The squall-line appears as a collection of echoes of irregular shape. These may or may not be interconnected. Each thunderstorm, in turn, may consist of several complex units or of single cells, the structures of which vary during the life-time of the storm as a whole. A plan view of a typical multicellular thunderstorm is shown in Fig. 7-3.

7.1.2 Development of a thunderstorm cell:

Byers and Braham (1949) have suggested that the life cycle of a thunderstorm cell can be divided into the following three stages (Fig. 7-4):

7.1.2.1 Cumulus stage:

The cumulus stage of the development of a cell is characterized by the presence of updrafts throughout. This causes the cell to grow upwards while entrainment of ambient drier air takes place through the sides. With the continued increase in updraft velocity, a large amount of moisture condenses and its weight exceeds the amount that can be supported by the updraft. Thus precipitation begins and a downdraft starts developing.

The updrafts were usually 1.2 to 1.5 km across and did not vary with height up to 7.5 km MSL. Maximum updraft velociti

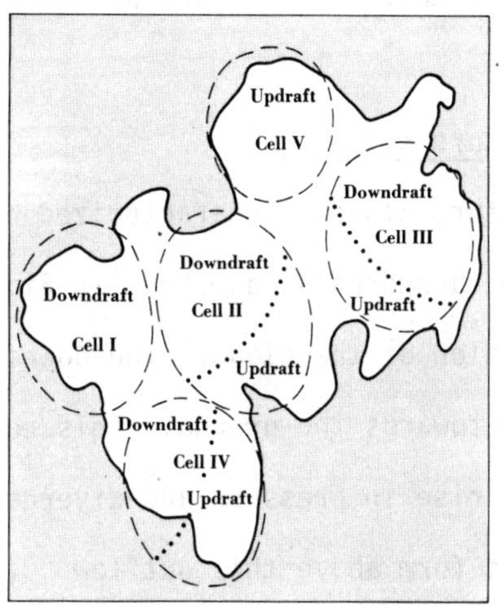

FIG. 7-3 Plan view of a typical multicellular thunderstorm. (from Petterssen, 1969)

varied between 5 and 15 m sec^{-1}. The updrafts were usually broad[illegible] and stronger than the downdrafts.

The growth of the radar echo top often proceeded in two or three steps of about 1.5 km each. The interval between successive turrets was about 18 minutes.

7.1.2.2 Mature stage:

The mature stage is characterized by the presence of both updrafts and downdrafts, as shown in Fig. 7-4, especially in the lower portion of the cloud. The downdraft brings cold air in the rain area towards the ground. This cold outflow is accompanied by a rise in pressure and divergence of surface winds. New cells form above this outflow. The regions of updrafts dominate throughout this stage and are warmer than the clear air environment at the same level.

7.1.2.3 Dissipating stage:

The cell enters the dissipating stage when the maximum downdraft velocity is reached at the expense of the updraft. The downdraft occupies the entire cell but eventually weakens and disappears. An individual convective cell has, on the average, a lifetime of half an hour, though a multicellular storm may persist for several hours.

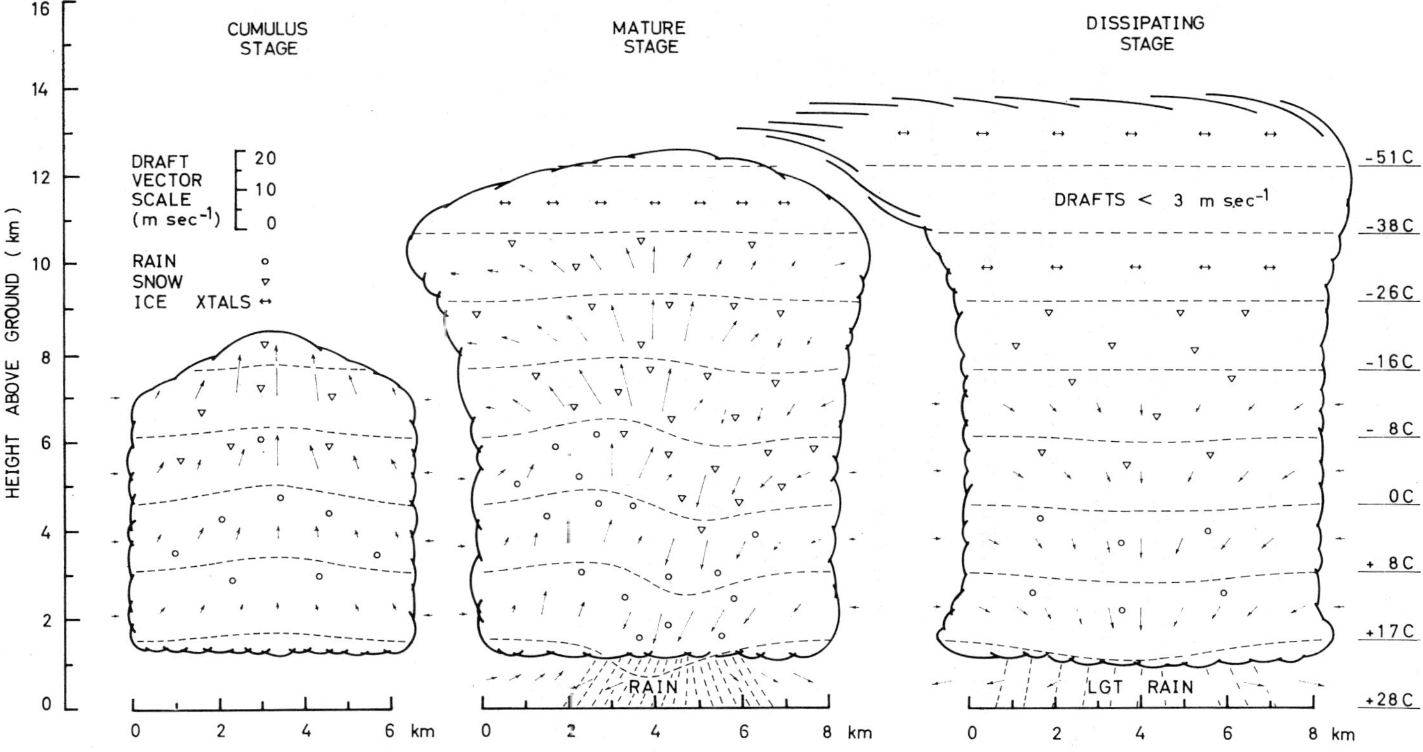

FIG. 7-4 (Figure legend on next page)

FIG. 7-4 The Byers-Braham thunderstorm model. In the cumulus stage the cloud is characterized by updraft throughout its entire depth; precipitation particles are already evident. The mature stage is marked by the onset of a precipitation-induced downdraft on the upwind side of the storm. An updraft still exists throughout the storm depth on the downwind side with maximum vertical velocities of 15 m sec^{-1}. Finally, in the dissipating stage, a diffuse anvil structure appears at the storm top and the downdraft spreads to occupy the major portion of the storm. (Adapted from Byers and Braham, 1949 by Chisholm and English, 1973)

7.1.3 Hail:

The Thunderstorm Project was conducted in regions where the hail frequency is low. Hence, it is not surprising that except for three days no hail was encountered in either Florida or Ohio thunderstorms. Hail was located most frequently at the middle levels of the clouds and was associated with particular cells within the storm. Four cells having updraft speeds greater than 15 m sec^{-1} produced hail. In three other cells having updraft speeds of less than 12 m sec^{-1}, hail development was absent.

Thus the Thunderstorm Project revealed many characteristics of air-mass type, small thunderstorms. It must be emphasized, however, that these storms are not similar in all respects to those larger and more intense steady-state storms occurring, for example, in the northern plains states of the USA.

7.2 Models of hailstone growth:

There seems to be a general agreement that hailstones grow predominantly by the accretion of supercooled water droplets. However, no really satisfactory quantitative theory of the growth of large hailstones exists which accounts for their multi-layered structure. Review of the past work on theories of hail growth indicates many divergent views. The broad concepts of important models are outlined below.

7.2.1 Multiple incursion model:

The formation of the opaque layers has generally been explained in this model in terms of collection of ice crystals or accretion of supercooled droplets in the colder region of the cloud; the transparent layers were attributed to the liquid coating obtained when the hailstone falls in the warmer region below 0°C level. The multiple incursion model, thus, pictures cyclical ascents and descents of the hailstones above-and-below the melting level of the cloud (Humphreys, 1940). The resulting alternate opaque and clear layers were then accounted for by postulating the number of oscillations of a hailstone above and below the 0°C level due to a pulsating updraft. There are two stringent requirements of this model: first, the strong updraft has to pulsate across the melting level and, second, it must do so in the lower region of the cloud which extends in the vertical to a considerable altitude. Thus, the growth of hailstones 5 cm in diameter implies pulsating updrafts with velocities greater than 30 m sec^{-1} across the 0°C level in the lower region of the cloud. Such conditions are never found. Besides, the clear layers of ice often observed in a hailstone are too thick to have been formed by the freezing of thin films of water. For these reasons, the multiple incursion model has been discarded.

7.2.2 Schumann's theory of hailstone growth:

The first fairly detailed mathematical analysis and quantitative treatment of hailstone growth was presented by Schumann (1938). In his calculations, the growth of the hailstone embryo is ignored and its history is followed only after it has attained a diameter of 0.5 cm. Schumann assumed that a hailstone grew spherical by sweeping up all the supercooled water droplets lying in its fall path. The radius of the hailstone was determined in terms of the distance traveled with respect to the surrounding air, liquid water content and updraft velocity. In time dt, let the hailstone move through a distance dS and sweep through a volume $\pi r^2_H dS$ where r_H is the radius of the hailstone. Thus, it acquires an added volume of $w\pi r^2_H ds/\sigma$ where σ is the average density of hailstone and w is the concentration of liquid water. This added volume is equal to the surface area times the small increase in radius, i.e., $4\pi r^2_H dr_H$. Therefore

$$wdS = 4\sigma dr_H \tag{7-1}$$

Integration between the proper limits gives

$$wS = \sigma(4R_H - 1) \tag{7-2}$$

where R_H is the ultimate radius of the hailstone. The initial radius is 0.25 cm. He calculates the distance (S) which the hailstone must travel to attain a radius of 4 cm, assuming w equal to 5 g m^{-3} and σ = 0.6 g cm^{-3} which results in a value of S of 16 km.

The time required for the growth of the hailstone in terms of its radius was calculated with the help of equation (7-1).

$$4\sigma \frac{dr_H}{dt} = w \frac{ds}{dt} = w\mathcal{V}_H \qquad (7\text{-}3)$$

where $\mathcal{V}_H$ is the terminal velocity of the hailstone. Assuming the dependence of the "drag coefficient" on Reynolds number as calculated by Bilham and Relf (1937), the relation

$$\mathcal{V}_H = 2000(\sigma r_H/\rho_{air})^{\frac{1}{2}} \qquad (7\text{-}4)$$

was used by Schumann to calculate the distance traveled by the hailstone relative to a fixed level. The actual distance traveled is equal to the distance traveled (S) with respect to the air, minus the upward distance traveled by the air itself.

$$S - Ut = \frac{\sigma}{w}(4R_H - 1) - \frac{U(\sigma \times \rho_{air})^{\frac{1}{2}}\ (4R_H^{\frac{1}{2}} - 2)}{1000w} \qquad (7\text{-}5)$$

where U is the vertical upward velocity of the air in cm/sec.

From these calculations, Schumann concluded that the principal factors which determine the ultimate size of a hailstone are the height at which its nucleus is formed, its average density, the average upward velocity of the air and the concentration of the condensed water in the region of the cloud where the temperature is below 0°C. According to these calculations a hailstone of 2.2 cm radius could fall from a cloud of liquid water content 4 g m^{-3} and an average updraft of only 15 m sec^{-1}.

When these equations are used to calculate hailstone growth in a model cloud with adiabatic properties and a steady updraft, the largest hailstones which can arise by continuous growth have radii of about 2 cm at the 0°C level. Such size is considerably less than that of hailstones with radii of up to 5 cm or higher which are frequently experienced at the ground. The discrepancy is increased, though only slightly, if we take into account the melting of stones, because the shrinkage is negligible if the radius considerably exceeds 1 cm.

Although he assumed in the beginning that all the water collected by the hailstone would freeze on its surface, Schumann was able to show in the latter part of his paper that this may not be true, even with air temperatures below 0°C; the rate of growth of a hailstone depends among other things on the rate at which it can dissipate the latent heat of fusion. He, therefore, determined critical values of the liquid water content for given values of the ambient temperature. The growth of a hailstone would be even more restricted at higher ambient temperatures, since the water captured by the stone cannot all freeze and some water may be shed as droplets. Although Schumann's theoretical treatment of the heat balance is correct, Ludlam's experimentally determined values have shown that his theoretical estimates of the transfer coefficients for heat and water vapor are not accurate. Ludlam (1950), however, came to the same conclusion as Schumann that hailstones might not be able to freeze all the water collected.

7.2.3 Ludlam's calculations of heat balance and hailstone growth:

Within the cores of large cumulus clouds air masses ascend at a high speed and cool at about the saturated adiabatic rate with respect to liquid water (Ludlam, 1950). Therefore, the liquid water content at any level within the core of such clouds can easily be read from an aerological diagram, assuming no precipitatio or storage of water in excess of adiabatic cooling. Figure 7- 5 shows the values calculated by Ludlam for clouds with bases at the 900 mb level which increase with the temperature at cloud base. The liquid water content above the melting level varies from about 1 g m^{-3} with -5°C base temperature to about 7 g m^{-3} when the base temperature is as warm as 25°C.

Ludlam assumed that a hailstone grows by accretion and the freezing of water accumulated on its surface. If this water freezes completely, let the rate of release of heat be H_1 and let H_2 be the rate at which heat will be transferred by conduction and forced convection. Furthermore, let H_3 be the rate of heat transfer due to the evaporation of water from the surface of the hailstone. The heat balance of the hailstone is obtained by equating the values for H_1 and $(H_2 + H_3)$. It gives us the critical value of liquid water content (w_c) for which the rate of heat loss is just sufficient to freeze all the collected water and to maintain the surface temperature of the hailstone equal to 0°C.

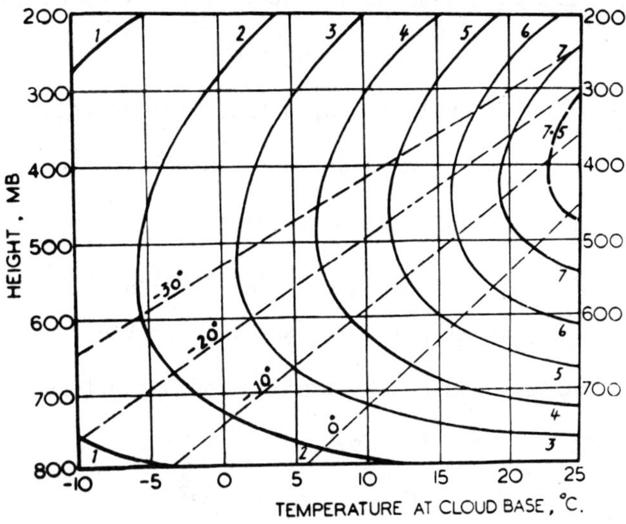

FIG. 7- 5 Liquid water content of cumuliform clouds. The values are calculated for clouds having bases at 900 mb; saturation with respect to liquid water is assumed throughout and the isopleths show the concentration of liquid water in g m^{-3}. The dashed lines indicate the temperature within the clouds. (from Ludlam, 1950)

Let us now write in detail these theoretical formulations

$$H_1 = H_2 + H_3 \qquad (7\text{-}6)$$

The mass of water collected by a spherical hailstone per unit time will be $E\pi r^2{}_H \mathcal{V}_H w$. (Terms are defined below) Hence, substituting proper values for rates of heat transfer in (7-6), we get,

$$\begin{aligned} &E\pi r^2{}_H \mathcal{V}_H w \;\{L_f + C_w(T_a - T_o) + C_i(T_o - T_H)\} \\ &= 2\pi K r_H (T_H - T_a)\;(Nu) + L_v 2\pi r_H D\Delta\rho_v (Sh). \end{aligned} \qquad (7\text{-}7)$$

After proper substitution for Nusselt number (Nu) and Sherwood number (Sh) as,

$$Nu = Sh = \chi Re^{\frac{1}{2}} \qquad (7\text{-}8)$$

in equation (7-7), we get the critical value for cloud water (w_c) for which the rate of heat loss is just that required for all the collected water to freeze; the temperature of the surface of the hailstone will be 0°C. χ in equation (7-8) is equal to 0.6. (See Ranz and Marshall, 1952; Thorpe and Mason, 1966).

$$w_c = \frac{2\chi Re^{\frac{1}{2}}}{E r_H \mathcal{V}_H} \; \frac{[K(T_H - T_a) + L_v D\Delta\rho_v]}{[C_w(T_a - T_o) + L_f]} \qquad (7\text{-}9)$$

Substituting for $Re = 2\mathcal{V}_H r_H \rho_a/\mu$, the condition for the stone to become wet is

$$r_H \mathcal{V}_H > \frac{8\chi^2\rho_a}{\mu E^2 w^2} \left\{ \frac{K(T_H - T_a) + L_v D\Delta\rho_v}{C_w(T_a - T_o) + L_f} \right\}^2 = f_2(z) \qquad (7\text{-}10)$$

where,

E	is the average collection efficiency;
r_H	is the radius of the spherical hailstone;
w	is the concentration of liquid water;

υ_H	is the fall velocity of the hailstone relative to the cloud droplets.
L_f, L_v	are the latent heat of fusion and evaporation, respectively;
C_w, C_i	are the specific heats of water and ice, respectively;
T_o	is 0°C;
T_a	is the ambient air temperature;
T_H	is the surface temperature of the hailstone;
K	is the thermal conductivity of air;
μ	is the dynamic viscosity of air;
D	is the coefficient of diffusion of water vapor in air;
$\Delta\rho_v$	is the difference between vapor concentration at the surface of the sphere and that in the remote environment;
ρ_a	is the air density;
Nu	is equal to $2.0 + 0.6\ Pr^{1/3}\ Re^{1/2}$;
Pr	is the Prandtl number and is equal to $\mu c_p/K$;
c_p	is the specific heat of air;
Sh	is equal to $2.0 + 0.6\ Sc^{1/3}\ Re^{1/2}$;
Sc	is the Schmidt number and is equal to $\mu/\rho_a D$.

If the concentration of cloud water is $< w_c$, for a hailstone of given radius and for given ambient conditions all the accreted water is frozen, as the temperature of water is below 0°C. The dry growth rate under these conditions is given by

$$\frac{dr_H}{dt} = \frac{E\upsilon_H w}{4\rho_i} \tag{7-11}$$

when ρ_i is the density of deposited ice.

However, when $w > w_c$, more water is collected than can be frozen. Ludlam assumes that the excess water under these conditions accumulates as liquid film, and part of it is shed. This type of growth would be 'wet' growth. The rate of deposition of transparent ice beneath the liquid film is given by

$$\frac{dr_H}{dt} = \frac{2\pi r_H \chi Re^{\frac{1}{2}} \{K(T_H - T_a) + L_v D\Delta\rho_v\}}{4\pi r_H^2 \rho_i \{C_w(T_a - T_o) + L_f\}} \qquad (7\text{-}12)$$

Neglecting the change of mass due to evaporation and since $Re = 2\nu_H r_H \rho_a/\mu$, equation (7-12) becomes,

$$\left(\frac{r_H}{\nu_H}\right)^{\frac{1}{2}} \frac{dr_H}{dt} = \left(\frac{2\rho_a}{\mu}\right)^{\frac{1}{2}} \frac{\chi}{2\rho_i} \left\{\left[\frac{K(T_H - T_a) + L_v D\Delta\rho_v}{C_w(T_a - T_o) + L_f}\right]\right\} \qquad (7\text{-}13)$$

$$= f_3(z)$$

where $\rho_i = 0.9$ g cm^{-3} and χ is a constant as stated earlier.

Values of $f_2(z)$ and $f_3(z)$ have been calcualted by Ludlam (1958) with $\chi = 0.6$ and, therefore, are applicable to smooth spheres. These values are given in Table 7-1.

The critical value of w for droplets of different sizes at ambient temperatures between 0°C and -30°C, is shown in Fig. 7-6. The model cloud has a base at 900 mb and base temperature of 10°C. The actual value of w, calculated from the adiabatic ascent from the cloud base, assuming no mixing, is shown by the dashed line transferred from Fig. 7-5. If it exceeds the critical value for an embryo of given radius then liquid water will accumulate on the embryo. The dashed line, which may be called the dry and wet growth limit, therefore divides the graph into two parts: to the right, all water collected on the embryo freezes solid; to the left, the embryos acquire

TABLE 7-1

Parameters $f_2(z)$ and $f_3(z)$ as functions of height, z, for a smooth spherical hailstone following an adiabatic ascent above a cloud base at 900 mb and 20°C. (from Ludlam, 1958)

Z km	$10^{-2}f_2(z)$ c.g.s. units	$f_3(z)$ c.g.s. units
6	0.1	0.1
7	0.6	1.0
8	1.8	3.0
9	4.2	6.1
10	9.5	10.6
11	20.0	16.4

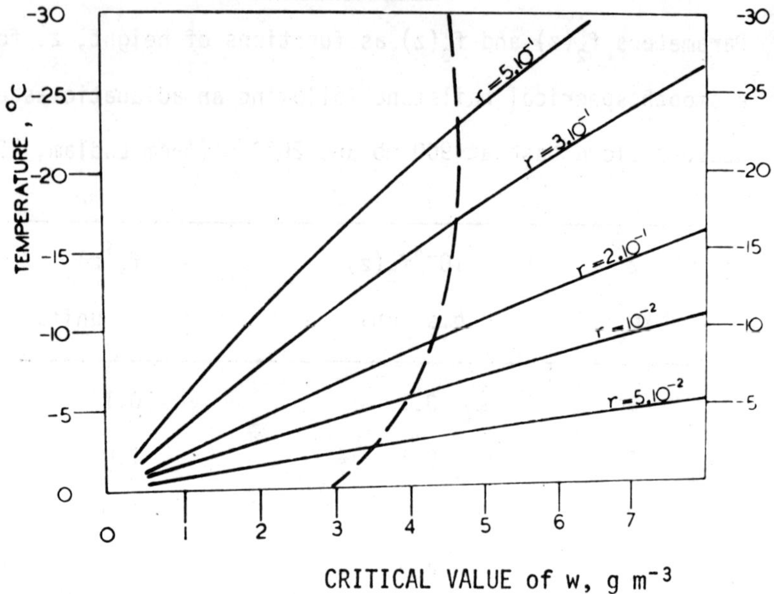

FIG. 7- 6 Critical liquid water concentration for coagulation-elements of various radii in cloud with base at 900 mb and base temperature of 10°. If at a particular temperature the actual concentration [shown by the dashed lines] exceeds the critical value, then liquid water accumulates on the coagulation-element. (from Ludlam, 1950)

liquid skins.

Ludlam (1958) assumed a model cloud whose base is at 900 mb and 20°C. The updraft in the supercooled region was assumed to be constant in height and time. The calculations of hailstone growth were made with the help of the growth equations (7-11) and (7-13), assuming that the embryo originates with the freezing of a small drop between the temperature range of -5°C and -10°C.

The growth of two droplets, frozen at -5 and -10°C, respectively, in a steady updraft of 20 m/sec is depicted in Fig. 7-7. The embryos become wet at about 10 and 11 km, respectively, and they finally attain radii at the 0°C level of about 1.25 and 1.5 cm. This is about the maximum size attainable with the steady updraft of 20 m sec^{-1}. If the updraft is increased, for example, to 23 m sec^{-1} to increase the growth time, the stones are lifted above the -40°C level into a region where rapid general freezing of the cloud water and practical cessation of growth would occur. The maximum diameters attained by these hailstones in the assumed cloud model are small as compared with the values of 5 to 8 cm reported for large hailstones. Moreover, if a model cloud has a lower base temperature than 20°C and, therefore, a somewhat lower concentration of liquid water in the supercooled region, it would further restrict the growth of the embryos. Ludlam points out that a more

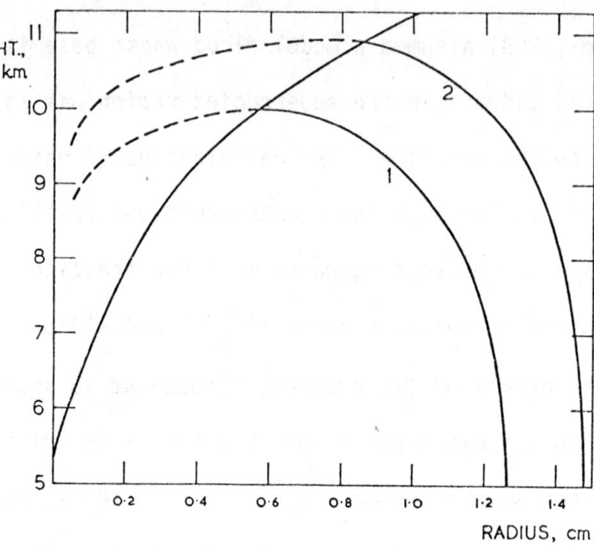

FIG. 7-7 Growth of hailstones in the model cloud containing the adiabatic liquid water content and with base at 900 mb, 20°C. The curves marked (1) and (2) show the growth with U = constant = 20 m sec^{-1} following the freezing of a small cloud droplet at -5° and -10°C, respectively. Curve (3) shows the conditions under which a hailstone becomes wet having a density of 0.9 g cm^{-3}. (from Ludlam, 1958)

favorable condition for maximum growth seems to be a larger initial size of the embryo, such as a raindrop, rather than a small cloud droplet.

7.2.4 Spongy accretion:

Although studies by Schumann (1938) and Ludlam (1950, 1958), as discussed in the earlier sections, did not produce a satisfactory theory, they led to the formation of a set of relations of the growth rate and heat balance of the hailstone. In these studies, the notion of wet growth of hail was advanced on the assumption that the unfrozen water was shed in the wake of the growing stone; this shedding restricted the hailstone growth. However, the laboratory work of List (1961a) and Macklin (1961) has shown the inaccuracy of this assumption. With the help of the studies of spongy accretion in high liquid water content, both List and Macklin have shown that some of the unfrozen water in wet growth is retained within the ice structure.

Taking into account spongy accretion, different formulations of the problems of heat balance and hailstone growth have been attempted by List (1963a), Macklin (1963), List, Schuepp and Methot (1965) and List and Dussault (1967). Some of the excess water is likely to be retained in a framework of spongy ice and, hence, limitations on the growth imposed in Ludlam's calculations are not realistic.

To obtain realistic hailstone growth, as observed in nature, various models suggest either keeping the embryo and then the hailstone in the cloud for a sufficiently long period by recycling it, or assuming that it grows in very high liquid water content. The features of these models will now be discusse

Formerly, there was common belief that wind shear at upper levels impeded the formation of thunderstorms. However, later research has shown that the presence of shear helps in organization of the storm. Weickmann (1953) has pointed out that a storm within a sheared environment is in motion relative to different winds at different levels and sweeps up moist air as it migrates through it. The models proposed by Dessens (1960) and Ludlam (1961) and the studies reported by Marwitz (1972a), ar typical in emphasizing the effect of wind shear at upper levels on hailstorm development.

The photographs taken by Henderson (1972) of a large hail producing air mass thunderstorm are shown in Fig. 7- 8 (a) and (b). Anvil is often observed to propagate outward in all directions from the primary inflow area which is along the traili edge. On the other hand the third photograph (Fig. 7- 8 c) shows the inflow area along the frontal edge of a squall line in Northeast Colorado. This is a classic example of inflow area related to squall line inasmuch as the inflow is easy to identify and the difference between characteristics of this type of storm and air mass type is clearly illustrated.

7.2.5 F. H. Ludlam's model:

Ludlam (1961) has made estimates of the distribution of temperature and cloud water concentration with respect to height in a cumulonimbus which are shown in Table 7-2.

FIG. 7-8(a) Legend on next page.

FIG. 7-8 (a) View looking north at giant air mass cumulus development moving northeast. Inflow on trailing edge of system shows two important updraft areas. [By courtesy of Mr. Thomas Henderson, Atmospherics Incorporated, Fresno, California]

FIG. 7-8(b) Legend on next page.

FIG. 7-8 (b) Close up view shows a single major inflow area to large hail producing air mass thunderstorm. Anvil is often observed to propagate outward in all directions from the primary inflow areas. [By courtesy of Mr. Thomas Henderson, Atmospherics Incorporated, Fresno, California]

FIG.7-8(c) Legend on next page.

FIG. 7-8 (c) The inflow area is along the frontal edge of a squall line in Northeast Colorado. [By courtesy of Mr. Thomas Henderson, Atmospherics Incorporated, Fresno, California]

TABLE 7-2

Typical conditions inside a severe storm cloud.
(from Ludlam, 1961)

Levels	Height km	Cloud temperature °C	Cloud-water concentration $g\ m^{-3}$	Proportion of water frozen	Updraft speed $m\ sec^{-1}$
Cloud top	14	-80	-	all	0
Tropopause	$11\frac{1}{2}$	-55	2	all	50
Top of supercooled zone	10	-40	4	all	50
	9	-30	4	small fraction	45
	6	-10	4	one-millionth	30
Base of supercooled zone	4	0	3	none	20
Cloud base	2	10	0	-	10
Ground	0	25	-	-	0

He assumes an updraft of increasing speed between the 0°C isotherm and the -55°C level. Ludlam further suggests that the updraft is not upright but is strongly tilted due to wind shear in the upper levels. He arrived at this conclusion from intense study of the Wokingham Storm for which Browning and Ludlam (1962) tried to construct the general airflow pattern and devised a general model for the severe storms.

After initial multicellular development, the cloud changed drastically into an apparently steady-state configuration. Ludlam suggests that the airflow within the storm possesses updrafts and precipitation-maintained downdrafts, fed persistently from opposite sides. The radar structure showed the principal aspects such as high reflectivity region, 'wall,' 'hook' and 'echo-free vault.' The findings were similar to those described by Donaldson (1962) and Marwitz (1972b). The vault, they suggest, contains only small droplets, and large droplets are held in suspension above this region. In addition it is the region of the main updraft which feeds the storm. A three-dimensional model of the Wokingham Storm showing relative streamlines or air trajectories is shown in Fig. 7-

All the liquid water in this model is assumed to be in small unfrozen droplets of very small terminal speeds. In order to explain the growth of large hail it is also assumed that the embryos are as large as raindrops and are present at the 0°C isotherm. However, there is serious difficulty in explaining how such millimeter size drops can appear at low levels in a cloud havi

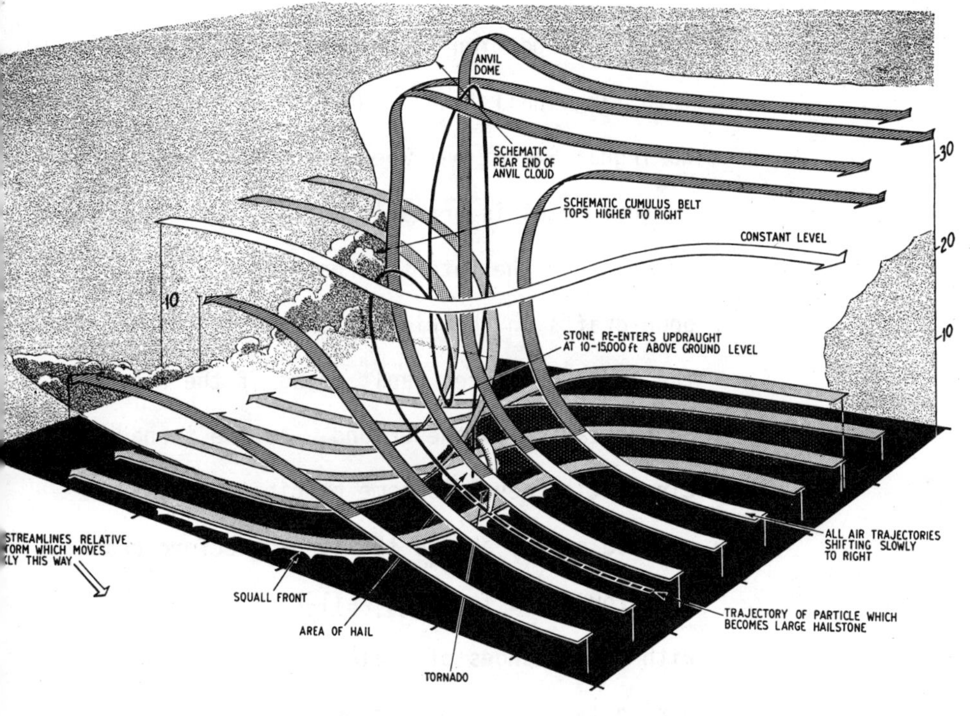

FIG. 7- 9 Three-dimensional model of the Wokingham Storm, showing relative streamlines or air trajectories (hatched where condensation has occurred). This model is tentatively regarded as representative of the severe traveling storms of middle latitudes. Hail reaches the ground in the white area and rain over the stippled area. A belt of cumulus is shown schematically lying in a wing over the squall front on the right flank of the storm. Precipitation forms in the larger cumulus; some particles are carried at levels above 13,000 ft forward and across relative

(continued next page)

FIG. 7-9 continued

to the storm and fail to enter the strong updraft. A small proportion rise to great heights and grow into large hailstones. No tornado accompanied the Wokingham Storm, but when one occurs its position in this model is probably where a funnel cloud is drawn, near the interface between the up- and down-drafts on the right flank of the storm, and in the radar echo-free vault ahead of the region of precipitation to the ground. As the tornado circulation develops some precipitation may be carried partly around the tornado before reaching the ground, to form the well-known 'hook' echo. With only changes of scale, especially the horizontal, the flow pattern represents also the large-scale baroclinic disturbance, and the tornado may even be regarded as in some ways analogous to the cyclone. (from Ludlam, 1963)

a very strong updraft.

The growth of a hail embryo and corresponding increase in its fallspeed take place rather slowly in its early stages. Therefore, such a particle would be carried up in a strong updraft in the supercooled region rather quickly. Ludlam, therefore, suggested that as the updraft is tilted, the hailstone is blown ahead in the tilted updraft, falls out of it and is picked up at the lower level, and continues its growth on a second journey. This process is assumed probable because the updraft is inclined to the vertical near the ground due to the inflow of air in the front portion of the storm.

It is further suggested that the small fraction of the re-entering particles of favorable size may be lifted slowly in the updraft. They keep growing at such a rate that the increase in their fall velocity is equal to the increase in the updraft speed, a very stringent requirement of this model. Finally, the hailstones acquire fallspeeds greater than the speed of the updraft; they move forward in the tilted updraft and fall downwards. The trajectory of the hailstone which grows in this fashion is shown in Fig. 7- 9. Particles beginning as initial small embryos will grow into small hail pellets and may, perhaps, fall out and melt before reaching the ground. But the large embryos of raindrop sizes may follow the path indicated earlier and develop into very large hailstones.

Browning, Ludlam and Macklin (1963) have made quantitative calculations for the growth of the reentrant hailstone. These calculations show that if the fall velocity of a hailstone increases, keeping pace with the increase in the updraft velocity, the hailstone will reach a size that is practically the maximum that can be achieved in this type of updraft. They suggest that the critical nature of this condition always restricts the concentration of large stones to small values. It is further suggested that under these conditions rather small fluctuations in the updraft velocity, or the physical characteristics of the hailstones themselves, could cause them to oscillate between the wet and dry growth regimes. This is how the layers of differing structure and opacity may be formed. For further discussion on this topic, the reader may refer to Chapter 8, Section 8.8.

The objections that can be raised to Ludlam's (1961) model are the following:

1. The presence of wind shear and the resulting tilted updraft are essential for the recycling of hailstones suggested in this model. However, air mass type thunderstorms without the presence of wind shear have been reported to produce large hail (Ratner, 1961; Geotis, 1963). Besides, Ratner (1961) could not find any correlation between wind shear and hail occurrence. Moreover

Sulakvelidze (1967) reports that when strong wind shear was present, hail formation was considerably weakened or altogether stopped.

2. The assumption of strong updraft with increasing speed between the 0°C isotherm and the -55°C level may be valid only in cases of a very severe storm.

3. The increase in fall velocity of a large hailstone has to keep pace, for the duration of its growth, with the increase in the updraft velocity. This is an extremely critical requirement of the model.

4. The analysis of the isotopic composition of a large hailstone, as discussed in Chapter 9 (Fig. 9-5) indicates that it never left the core of the updraft **during its major growth and did not grow by recycling.**

5. The presence of millimeter size embryos near the 0°C level has been assumed without any explanation.

7.2.6 Dessens' model:

Dessens (1960) found from both physical and statistical studies that, at least in France, a strong vertical development of a cloud was a necessary, but not sufficient, condition for the formation of a hailstorm. In addition he observed that severe hailstorms were associated with very strong horizontal winds between the levels 6 and 12 km. He proposed a hailstorm model which suggests that the updraft chimney does not draw well unless a strong wind blows over it at high levels as shown in Fig. 7-10.

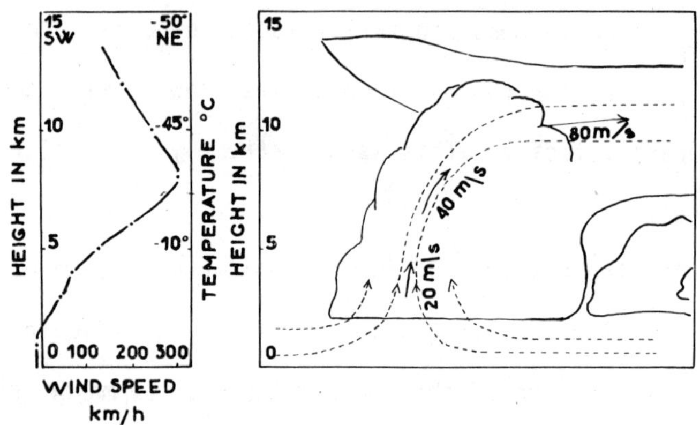

FIG. 7-10 A suggested structure for the hailstorm: [left] wind velocity versus height; [right] the chimney inclined to the vertical. (from Dessens, 1960)

The air in the core of the updraft gains energy from the instability in the lower layers and from the latent heat of condensation and of freezing. When mild wind shear exists in the atmosphere, the updrafts are approximately vertical. It is argued that if the air is not removed from the top, the updraft may not be able to draw the energy from the lower end. However, when there is strong horizontal wind present as shown in Fig. 7-10, a part of the air entrained in the updraft is sucked horizontally at the top, at a speed of the order of 80 m sec^{-1}. Dessens, therefore, suggested that a coupling takes place above 6 km height between the updraft and the very strong horizontal wind. This results in accelerating the updraft in a hailstorm. Thus, the updraft receives the kinetic energy of strong winds aloft in addition to the energy of thermodynamic origin at low levels, which prolongs its lifetime sufficiently to make possible the growth of large hailstones.

This model shows a correlation between the severity of a storm and the presence of wind shear at upper levels. However, it does not explain the hail growth process as such and the role of wind shear in it. The objections raised against Ludlam's model regarding wind shear (see Section 7.2.5) are also applicable to Dessens' model.

7.2.7 Hitschfeld and Douglas' model:

These two investigators argued that the updraft found in convective clouds is of the order of magnitude appropriate to the accumulation of rainwater rather than cloud water and hence, the possibility of rain accumulation to high water contents should not be overlooked. Hence, growth of particles was calculated for various updrafts and high liquid water contents. Hitschfeld and Douglas (1963) assumed an essentia steady updraft in the upper portion of the cloud with simple vertical stone trajectories. They calculated the spongy growth of stones in the high concentration of water up to 30 g m^{-3}. They assumed that a growing particle would retain most of the supercooled water it collected, even when its surface temperature is 0°C. One result of this assumption is a great increase in the hail growth rate.

The assumed vertical distribution of the updraft is critical to many parts of the theory. As no direct measurements or entirely convincing theory was available, they have therefore considered an assortment of likely updraft models (Fig. 7-11) of which C, E, and D have been used in their trajectory calculations. The growth of a hail embryo, originating as an ice particle 1 mm in diameter somewhere above the 0°C level, is calculated. The origin of such ice particles in a rain distribution is assumed to lie in the stochastic nature of the freezing events. The embryo is carried aloft by the updraft until its fall speed, increasing as a consequence of its growth, equals the updraft speed. Then, it falls back toward the ground. The composition of the particle is determined by computing the fraction of the

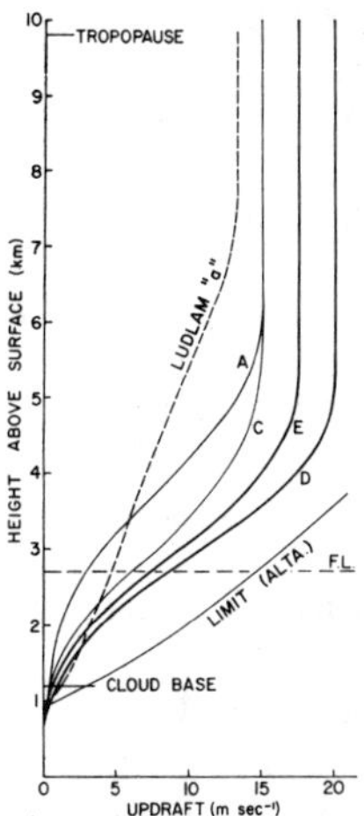

FIG. 7-11 Updraft models. 'LIMIT (ALTA)' is based on simple parcel theory applied to typical hail-day soundings, without regard to drag. LUDLAM'S 'a' involved the drag encountered by small bubbles. A and D are deduced from Battan's observations of first-echo development, as by Marshall and Gordon (1957). The computations to be presented were based on updraft models C, E, and D. (from Hitschfeld and Douglas, 1963)

liquid mass of the hailstone throughout its trajectory. The water in that region is supercooled and there is, therefore, a finite probability that any given drop will freeze. This probability is taken into consideration. The majority of drops of a given size group freezes at quite low temperatures; only a relatively small number of drops freeze at warmer temperatures. Thus 20 percent of the 1 mm diameter drops freeze at temperatures colder than -15°C, but only a small fraction of one percent freeze at temperatures warmer than -4.5°C; in a typical rain distribution this would still be about one per m^3. In their model, therefore, they picture these first frozen raindrops servir as the hail embryos.

With these assumptions, they have constructed a series of stone trajectories for updraft models D, E, and C (Fig. 7-12), stored-rain concentrations 10, 20 and 30 g m^{-3}, and an initial height of hail growth (where the embryo has 1 mm diameter) 3.4, 4.2 and 5 km above the surface, corresponding to temperatures of -5, -10 and -15°C. Such trajectories, 27 of them, are shown in Fig. 7-12. Time, in 2-minute intervals, is marked along the trajectories by large open circles; the fraction of the stone whic is liquid is marked, in steps of 10 percent, by black dots.

In twelve of these 27 sets, the updraft carries the embryo too quickly into the icy upper layers, where its growth is too slow and hence it does not overcome the updraft.

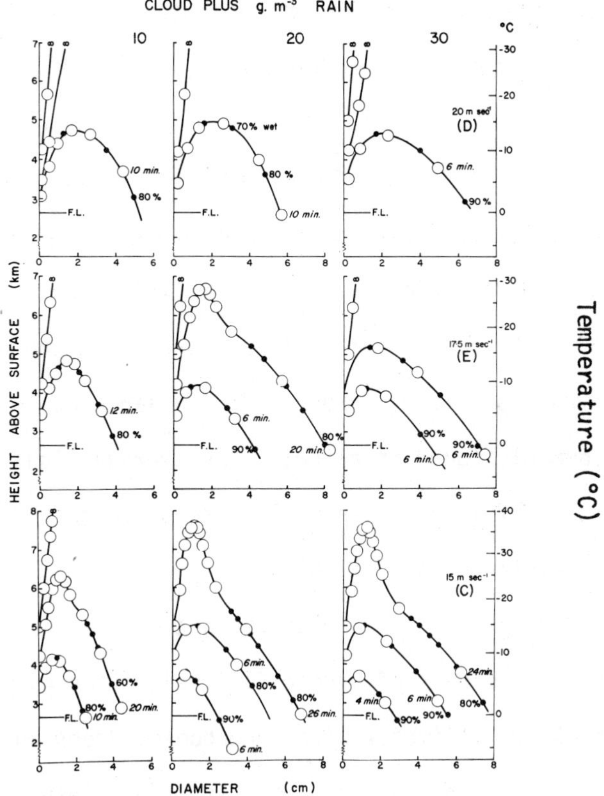

FIG. 7-12 Growth of the particles in the vertical. Each row of trajectories is for stated updraft model, each column for stated rain density. Big open circles along curves indicate time [in two-minute intervals] from start; black dots indicate [in steps of 10 percent] the liquid fraction of the particle. (from Hitschfeld and Douglas, 1963)

The remaining 15 trajectories do lead, however, to substantial growth, and produce particles which contain liquid fractions of 80 to 90 percent by the time they reach the 0°C isotherm. Only in nine of these cases, at some earlier stage, sizeable hail with liquid fractions 60 percent or less is produced. Thus, the hailstones grown in such high water contents, according to Hitschfeld and Douglas' calculations, would contain 60 to 90 percent of water and arrive at the ground as soft, mushy particles. Such soft and large hailstones are not commonly found in nature. The average liquid water content of large hailstones is reported to be 4.2 percent by weight (Gitlin et al., 1968).

Hitschfeld and Douglas did not discuss in detail how large hail embryos of higher fall velocities will be able to pick up millimeter-size supercooled drops and grow in high liquid water content. No attempt has been made to account for the layered internal structure due to wet and dry growths. Moreover, the assumed non-divergent updraft profiles in their model seem to be unrealistic. Furthermore, the real problem with their results was the excessive degree of wetness of the stones.

7.2.8 Model proposed by Sulakvelidze and his associates:

The concepts of this model of hail and precipitation formation were first developed at the High Altitude Geophysics Institute (known as VGI), Leningrad, USSR, in 1958. Some details of the proposed mechanism were later amended. However, the basic concepts have not changed since their formulation. Two important suggestions of the model are the parabolic shape of the updraft profile and the development of an accumulation zone.

7.2.8.1 Mapping the updraft profile:

Sulakvelidze et al. (1965, 1967) attempted to determine the vertical profile of the wind field in a developing cumulus and a cumulus congestus. This mapping of the updraft velocities was carried out by using radar to track small balloons fitted with corner reflectors. In most cases, the updraft velocity was observed to increase with height to a maximum value and then to decrease (Fig. 7-13). They have pointed out that the number of cases over which the averaging was carried out is not the same for all heights. Hence, some caution must be exercised in drawing conclusions about the exact values; however, the shape of the updraft profile should be accurate.

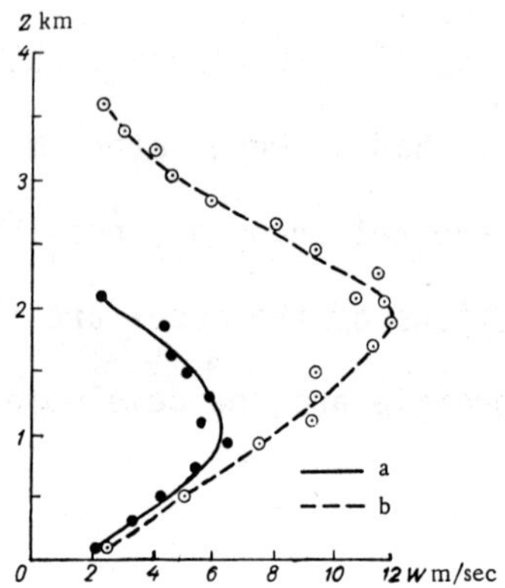

FIG. 7-13 Distribution with height above cloud base of averaged ascending current velocities in clouds. (a) - Cu and Cu med; (b) - Cu cong and Cb. (from Sulakvelidze et al., 1965)

The magnitude of the updraft velocity in the lower part of the cloud is represented by the equation

$$U_{(z)} = U_o + (U_{max} - U_o)\frac{Z}{Z_{max}} \qquad (7\text{-}14)$$

where $U_{(z)}$ is the updraft velocity at height Z, U_o is the velocity at cloud base where Z = 0, U_{max} is the maximum updraft velocity at the level Z_{max}.

In the upper part of the cloud where $Z \geq Z_{max}$

$$U_{(z)} = U_{max}\left(1 - \frac{Z - Z_{max}}{Z_a - Z_{max}}\right) \qquad (7\text{-}15)$$

where Z_a is the height at which U is zero.

The above equations describe fairly well the distribution of the updraft velocity in a developed cumulus. The level of maximum velocity is situated in the middle of the cloud.

The shape of the updraft profile obtained in the case of a Cb is similar to the one in a developed cumulus (Fig. 7-13). The vertical velocity is an almost linear function of height above cloud base up to a certain maximum, and then it decreases towards the top of the cloud. The updraft velocities were 2 to 3 m sec^{-1}, at the cloud base, with a maximum value of 12 m sec^{-1}. The updraft velocity reached as much as 30 m sec^{-1} in very large mature thunderstorms. At times, both ascending and descending currents were observed simultaneously.

The temperature at the maximum updraft level was found to be a critical factor in determining the type of resulting precipitation. Thus, two prime factors related to updraft for hail formation were the value of the maximum updraft velocity and the temperature at which this maximum occurs.

7.2.8.2 The accumulation zone:

The other interesting suggestion of this model is the formation of an accumulation zone of hydrometeors. Water droplets carried into the cloud by the updraft coalesce with other droplets, and grow larger as they rise to higher levels. But, once past the level of maximum velocity as the updraft gradually decreases toward the top of the cloud, each growing droplet will rise to a level at which the updraft is just sufficient to support it. As the droplet continues to grow, it must fall to a level of greater updraft velocity; it would continue to grow and fall in the upper part of the updraft were it not for the fact that falling drops disintegrate when they reach a diameter of about 6 mm. Thus, if the updraft maximum is greater than 10 m sec^{-1} - the terminal veloci of a 6 mm drop - any droplet carried up into the region above the level of maximum updraft is trapped in a closed cycle of events: as the drop grows, it slowly falls and then shatters into fragments which, in turn, are carried upward and grow into larger droplets.....and so on. Under these conditions, an accumulation zone builds up above the level of maximum updraft. The liquid water content in this zone may be as high as 30 to 40 g m^{-3}, the average moist adiabatic values being 3 to 6 g m^{-3}.

The cross-sectional area of the accumulation zone at the radar reflectivity maximum was between 3 and 4 km^2; this agreed well with the area of the hailswath at the ground after due correction was made for the spreading effect due to wind. The volume of the accumulation zone was about 10 km^3 in most cases (with a maximum of, perhaps, 30 km^3).

7.2.8.3 The model:

As the embryo moves toward the top of the cloud, it is traveling at a speed depending on the magnitude of the updraft and the size of the embryo. On the way down, the embryo is falling slowly against an increasing updraft and spends a long time in a region of highly concentrated supercooled water (LWC = 10 to 40 g m^{-3}). Thus according to the model, the hailstone grows very slowly on the way up and very rapidly on the way down. The growth and fall of hail involve two stages of cloud particle growth. Drops in a hail cloud can grow up to 5.5 mm diameter. The freezing of water drops of this size occurs at relatively warm temperatures. Hail embryos, thus formed, grow by coalescence with large supercooled liquid drops.

Thus, the mechanism of the formation of rain and hail showers suggested by Sulakvelidze (1967) can be summarized as follows: hail formation is only possible when the accumulation zone is located between 0 and -25°C isotherms. Under these conditions, individual drops in the upper part of the accumulation zone freeze and act as hailstone embryos; these then grow by coalescence with droplets and large supercooled drops and descend to the central supercooled part of the accumulation zone, where rapid hail growth occurs by coalescence with large supercooled drops. The liquid water content of the large drop fraction is expected to be more than 20 g m^{-3}.

In almost every case for which strong wind shear was present, they report that there was little or no hail.

The main objections that can be raised to Sulakvelidze's model are the following:

1. Large hail embryos have higher fall velocities than millimeter-size supercooled drops. Hence, they will fall to a lower level and will not be able to grow further by accreting large supercooled drops in the accumulation zone, as suggested in this model.

2. The hailstones grown in such high liquid water content up to 40 g m^{-3} would contain a large amount of water (60 to 80 percent) as discussed in the previous section. Usually hailstones are hard and contain only about 4 percent water by weight (Gitlin et al., 1968).

3. The development of the internal structure of layers due to dry and wet growth cannot be accounted for. Most of the growth in high liquid water content would be wet.

7.2.9 Numerical modeling:

Recently there have been some attempts at numerical modeling of hailstone growth, but none of these models explains the internal structure of hailstones. Many unrealistic assumptions are made, such as a horizontally homogeneous and non-divergent updraft, injection of embryos at the freezing level without explaining the microphysics of hail embryos growth (List et al., 1968; Charlton and List, 1972), assumption of instantaneous mixing, unrealistic hail development and growth beneath the updraft maximum (Danielsen et al., 1972), unrealistic high liquid water content profiles and major hailstone growth below updraft maximum (Musil, 1970; Dennis and Musil, 1973), extensive parameterization of the microphysical processes and assuming hailstones grow by accretion of raindrops (Wisner et al., 1972). These shortcomings raise enough serious doubts as to the accuracy of such models.

CHAPTER 8

NEW MODEL OF HAILSTONE GROWTH

We have discussed in the previous chapter the models that have been proposed by different investigators to explain hailstone growth. One of the two main suggestions for the rapidity of growth was that it takes place in very high liquid water content (Hitschfeld and Douglas, 1963; Sulakvelidze et al., 1965; Sulakvelidze, 1967). The other suggestion was that the embryo and then the hailstone is recycled in the tilted updraft due to wind shear at upper levels (Ludlam, 1961) and hence the embryos get sufficient time to grow to larger sizes. The main objections to these ideas are also discussed in the previous chapter, and one must conclude, therefore, that none of these models explains adequately the mechanism of hail formation.

Recently a new model for the hailstone growth was suggested by the author (Gokhale and Rao, 1969a). The mechanism proposed in this model and the way it explains the hailstone growth are discussed in the pages to follow.

8.1 Structure and velocity of the updraft:

The computation of updraft velocities is difficult because of the unknown parameters of drag, friction, and the dimensions of the cell and inadequate knowledge of the dynamics of cloud formation. It is equally difficult to measure such speeds during aircraft penetrations. However, with the help of radar and balloon observations, updraft speeds of between 20 and 30 m sec^{-1} have been measured and updraft distribution has been observed to be maximum at middle levels and to decrease upward (Sulakvelidze et al., 1965; Simpson and Wiggert, 1969). Refer to Fig. 7-13 in Sec. 7.2.8 for the nature of the updraft profile reported by Sulakvelidze et al. (1965).

A detailed study of a cloud which developed a vigorous updraft and produced small hail has been made by Battan and Theiss (1966) by means of a vertically pointing pulsed-Doppler radar. The profiles of inferred vertical velocities of the air do show the maximum values reached in the upper part of the developed cloud and a decrease toward the top.

Packets of slow-falling (< 30 cm sec^{-1}) chaff were released at cloud base in the organized updraft of severe hailstorms by Marwitz (1971). They were tracked by M33 radar. The vertical velocity profiles from the chaff tracks obtained on 9 July 1971 are presented in Fig. 8-1. The vertical velocities reached a peak value 4 km above the cloud base and decreased upward, resulting in an updraft profile similar to the one reported by Sulakvelidze et al. (1965).

The NHRE dropsondes which were launched about 14 km above sea level gave vertical wind soundings. They floated in updrafts with little vertical motion for periods as long as 20

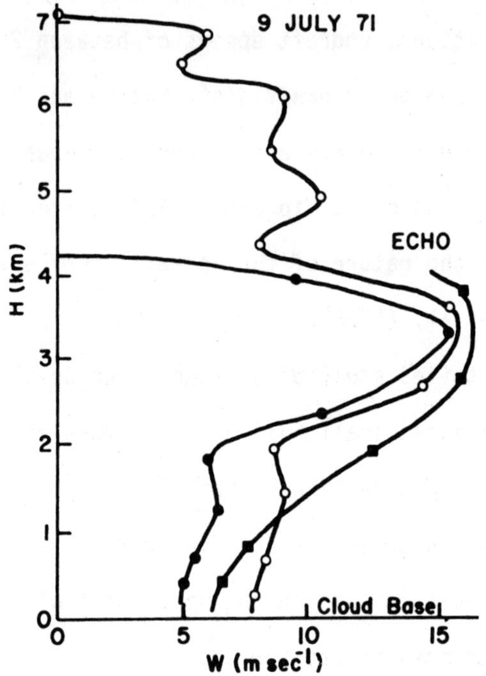

FIG. 8-1 Vertical velocity profiles for 9 July 1971. (from Marwitz, 1971).

minutes, implying velocities in excess of 30 m sec^{-1} (NCAR, 1972). Besides, in NHRE experiment vertical winds were measured by ten special dropsondes side by side over a distance of four km. The updrafts were up to 13 m sec^{-1} at -12 to -32C levels with much variability (Bushnell, 1973).

In the Thunderstorm Project (Byers and Braham, 1949) radar examination of the developing clouds indicated a cellular structure. A particular 'cell' giving a well-defined radar echo was a large cloud containing a number of smaller active units called 'turrets,' which represent individual columns of bubbles. Thus it was observed that the highest echoes were built in steps in which successive turrets extended about 1.5 km above the preceding one. Each cell passed through three distinct stages: the cumulus stage, the mature stage, and the dissipating stage, as described in Sec. 7.1.2.

In the summer of 1968, during 22 hail cases in South Dakota, it was reported (Dennis et al., 1970) that convective storms in the Rapid City area tend to be of the pulsating type rather than the steady state. However, more intense thunderstorms possess during their lifetime a steady state phase (Ludlam, 1958; Newton and Newton, 1959; Donaldson et al., 1960; Newton, 1963; Browning, 1963).

Three important characteristics of the structure of the updraft in a developing cumulonimbus are revealed by the observations mentioned above:

(i) Updraft velocity in the core of a cell increases with height, reaches a maximum value (in excess of 30 m sec $^{-1}$ in intense storms) and then decreases towards the top of the cloud.

(ii) The updraft normally has a pulsating nature and the updraft velocities are built up in steps by the rise of successive turrets.

(iii) A rapidly developing strong updraft may end up having a steady state phase for a limited period.

The updraft profiles and structure, as summarized above, are assumed in the core of a cell of the model cloud which is described in the next section. It is also assumed that the parameters such as updraft velocity, temperature, liquid water content are constant across the core of the cell which is characterized by little mixing (Ludlam, 1950). Thus a three dimensional development of medium and large hail in the core of the cell can be analyzed with the help of a one-dimensional model, taking into account the variation of different parameters in the vertical.

8.2 The model cloud:

8.2.1 Base height and temperature:

The model cloud shown in Fig. 8-2 has the typical characteristics of a well developed cell of a hailstorm. The base of the cloud is at 1.5 km with the temperature +10°C. The top extends above 10 km. The temperature distribution inside the cloud is assumed to be adiabatic with a moist-adiabatic lapse rate of 6.5°C km^{-1}. Thus, the melting level is at the 3 km height.

8.2.2 Fall velocities of particles:

The slant lines in Figs. 8-2 and 8-3 indicate the fall velocities of water drops and of hailstones respectively, at different elevations. These values are calculated using the Foote and Du Toit (1969) equation for determining terminal velocities aloft. Foote and Du Toit, using Davies' data (Davies, 19

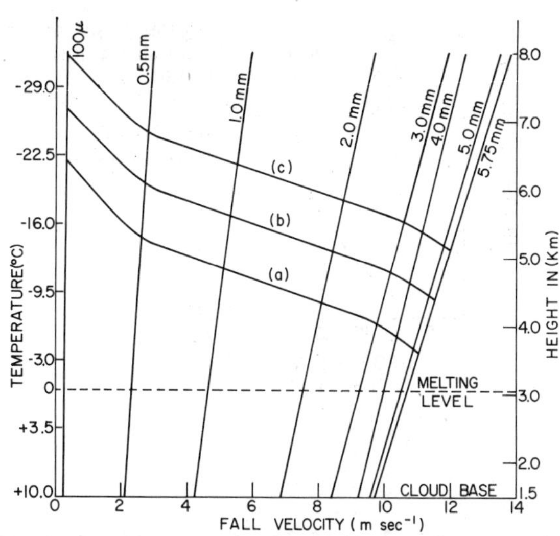

FIG. 8-2 Characteristics of the model cloud. The cloud base temperature is +10°C and height above ground is 1.5 km. The slant lines indicate the fall velocities of the indicated drop diameters at different heights. Curves (a), (b) and (c) indicate the freezing temperatures of 5 percent of the drops, 50 percent of drops and all drops, respectively. (from Gokhale and Rao, 1969a)

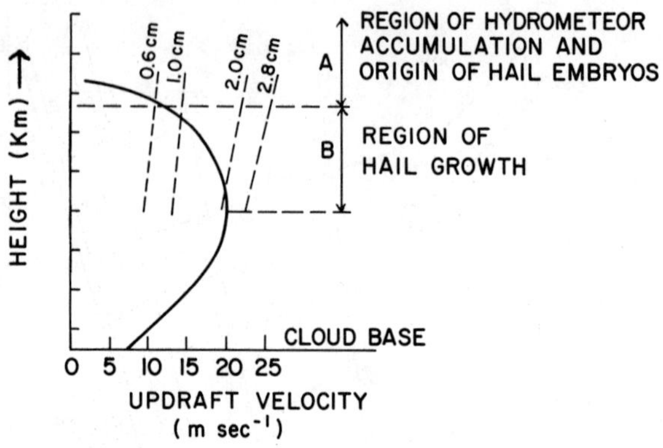

FIG. 8-3 Typical profile of a developed updraft. The fall velocities of different size hailstones are shown by dashed lines. The values along the Y-axis would depend upon how well the updraft has developed. (from Gokhale and Rao, 1969a)

derived the following general equation for the velocity of a particle of diameter (d) aloft as a function of density.

$$\mathcal{V}(d) = \mathcal{V}_0 (d)\, 10^Y \left[1 + 0.0023 \left\{1.1 - \rho/\rho_0 (T_0 - T)\right\}\right] \qquad (8\text{-}1)$$

where $Y = 0.43 \log_{10} (\rho_0/\rho) - 0.4 [\log_{10} (\rho_0/\rho)]^{2.5}$, and $\mathcal{V}(d)$ is terminal velocity of drop diameter (d), at air density ρ and temperature T. $\mathcal{V}_0$ (d) is its terminal velocity at air density ρ_0 and temperature T_0. The zero subscript refers to 20°C and 1013 mb pressure. The approximate form of their equation,

$$\mathcal{V}(d) = \mathcal{V}_0 (d) (\rho_0/\rho)^{0.4} \qquad (8\text{-}2)$$

is used to calculate the fall velocities aloft of drops as well as of hailstones. The values of $\mathcal{V}_0$ (d) used for water drops and for hailstones are those reported by Gunn and Kinzer (1949) and Macklin (1961) respectively.The terminal velocities aloft thus calculated agree reasonably well with those of Browning (1963).

8.2.3 Region of supercooled and frozen water drops:

For a population of supercooled droplets, each of volume V, the probability of freezing given by Bigg's (1953) equation has been expressed by Barklie and Gokhale (1959) in the following form:

$$- \ln N/N_0 = BV/\alpha a \; e^{aT_s} \qquad (8\text{-}3)$$

where N is the number of drops (originally N_0) remaining unfrozen at temperature T_s where $T_s = (T_0 - T)$ degrees Celsius and T_0 is

0°C. α is the cooling rate where $T_s = \alpha t$, t being the time. 'a' and 'B' are constants for the particular water sample.

From equation (8-3), one can derive

$$\ln V = Y - aT_m \qquad (8\text{-}4)$$

where Y is a constant and is equal to $[\ln \{ \ln (N_0/N) \} - \ln (B/\alpha a)]$ and $T_m = (T_0 - T)$ is the median freezing temperature, that is, the temperature when $N/N_0 = 1/2$. Thus equation (8-4) expresses the linea relationship between the logarithm of the volume of water drops and the degree of supercooling at which the drops freeze, and becomes a useful predictor of freezing events within a population of raindrops or cloud droplets balanced or ascending in an updraft.

The relative number of drops frozen as a function of temperature is calculated assuming a = 0.65 deg C^{-1}; the value of B appropriate to rain water, 10^{-4} cm^{-3} sec^{-1} (Barklie and Gokhale, 1959); and an average value of α equal to 2°C min^{-1}. It is based upon a rate of rise of the drops at 5 m sec^{-1} with a lapse rate of 6.5°C km^{-1}. The dependence of the median freezing temperature of supercooled water drops on the rate of cooling has been reported by Gokhale (1965a) over a wide range of cooling rates. For cooling rates between 0.3 and 5.0°C min^{-1} deeper supercooling is observed with increasing cooling rate; for cooling rates greater than 5.0°C min^{-1} the depth of supercooling becomes smaller. But, this dependence on cooling rates is rather small over the entire range. For example, the change in cooling rate by an order of magnitude will change the freezing temperature by less than two degre

Celsius. Figure 8-2 shows curves which indicate the freezing temperatures of 5 percent of the drops, 50 percent of the drops and all drops. The different sizes of the drops and their corresponding fallspeeds are indicated by the slanting lines which cross these curves. The temperatures are shown on the vertical axis. Thus, the curves (a) and (c) bracket the temperature range within which the freezing of various sized supercooled drops occurs. The warmest temperature is -4°C and the coldest temperature is -30°C for the drop diameter range of 5.75 mm and 0.1 mm.

8.2.4 Updraft profiles:

The most important of all the parameters is the updraft velocity and its variation with time. The structure and velocity of the updraft have been discussed in section (8.1). Though it is rather difficult to draw accurately the updraft profiles in any developing cloud, the typical shape of the profile of a well-developed updraft can be drawn as discussed by Squires and Turner (1962). They have presented a model of cumulonimbus updraft, which entrains environmental air according to the simple law that the inflow velocity at any height is proportional to the upward velocity of the plume. With this model, the shape of the cloud as well as its other properties follow from the dynamics, and the cloud development is size-dependent. Preliminary calculations result in realistic cloud shapes and reasonable values for the internal properties. The problems of convection dynamics and the growth of precipitation can hardly be separated. Squires and Turner have considered the effect on the cloud dynamics of changing the mass flux and arrived at updraft profile similar to the ones shown in Fig. 7-13, 8-1, and 8-3.

8.2.5 Liquid water content:

Four soundings taken in high-speed updrafts of severe thunderstorms indicate moist adiabatic ascent to mid-levels, implying that the cores of strong updrafts are undiluted by environmental air (Davies-Jones, 1974). The vertical distribution of the liquid water content in the core of the model cloud with a base height of 1.5 km and base temperature of +10°C is therefore calculated by assuming that the process of lifting from the base is adiabatic and is as shown in Fig. 8-4 (Ludlam, 1950). Variation in LWC in the region 4 km to 8 km where the major hailstone growth takes place (see section 8.5), is rather small as seen from Fig. 8-4; hence, in this region the average value of LWC can be assumed.

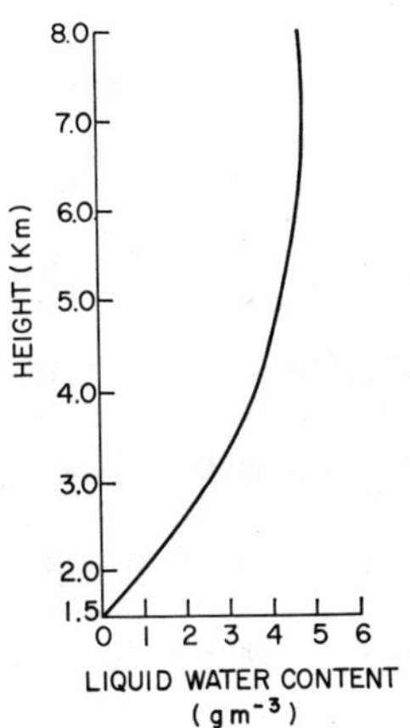

FIG. 8-4 Adiabatic values of liquid water content in the model cloud. (from Gokhale and Rao, 1969a)

8.3 The growth and accumulation of large hydrometeors in the upper part of a well-developed updraft:

8.3.1 The growth of millimeter size drops:

The location of the main region of particle growth is in the upper part of the updraft where its velocity decreases with height. The main growth does not seem possible in the lower part of the updraft as discussed in Sec. 8.3.2.1. The millimeter-size particles that are trapped in the upper part of the updraft cannot leave the updraft core unless they grow and their fall velocities become greater than the maximum velocity of the updraft, provided that no significant lateral motion occurs in the core. The growth of millimeter-size drops as they fall against the updraft is governed by the well-known coalescence equation. Figure 8-5 gives the growth in 50 minutes of a cloud droplet of 100 μ in a cloud containing liquid water content of 4 g m^{-3}. The size of the drop is calculated at one minute intervals. In fact, this curve could be used to estimate the growth of any drizzle or rain-size drop balanced in the updraft of the model cloud for a certain interval of time. For a growing drop which is falling continuously in an updraft, duration of growth and magnitude of the updraft maximum are the important factors which determine its final size. So the sizes of the drops that leave the core depend on how long the drops were balanced in an updraft core which has a certain liquid

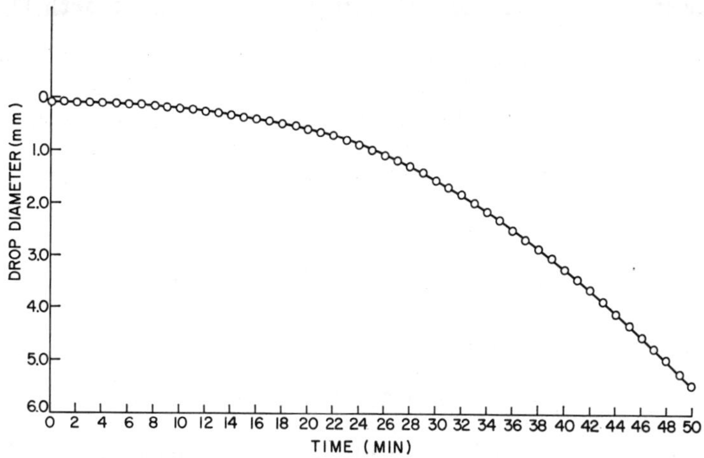

FIG. 8-5 Growth of a cloud droplet whose initial size is 100 μ. Liquid water content is 4 g m^{-3}. (from Gokhale and Rao, 1969a)

---

water content. If their resulting rate of increase of fall velocities due to their growth equals the rate of increase of the updraft speed, the drops would be balanced in that region for the maximum period, resulting in their maximum growth by coalescence and by Langmuir chain reaction.

Some drops, depending on their sizes and the ambient temperature, may freeze as discussed in Sec. 8.2.3. Such frozen drops will grow further by accretion.

8.3.2 Accumulation of hydrometeors:

There have been reports in the literature indicating that in thunderstorm clouds liquid water concentration as high as 30 g m^{-3} may develop (Marshall, 1961; Hitschfeld and Douglas, 1963; Sulakvelidze et al., 1965; Haman, 1967, 1968; Iribarne, 1968). Such concentration has also been invoked as discussed in section 6.2.1 to explain very high radar reflectivities observed aloft in convective clouds (Donaldson, 1961). Also from the rate of precipitation of sporadic severe cloud burst, high water accumulati have been derived. The evaporator, which measures the bulk water content of severe storms by converting it to vapor was recently flown on an armored aircraft as part of the National Hail Research Experiment (NHRE). The first such measurements ever reported indicate values upwards of 50 g kg^{-1} (NCAR, 1972).

Gokhale and Rao (1969b) have attempted to explain the development of regions of high liquid water content in a cumulonimbus with the help of a simple model and schematic diagrams. A summary of that development is given here. A time series of vertical velocity profiles is drawn through the core of the updraft. The liquid water is assumed to be distributed between both the small droplet fraction and the large droplet fract

Let us consider separately the growth and accumulation of millimeter-size drops in lower and upper parts of the updraft profile.

8.3.2.1 Growth and accumulation in lower updraft region:

The updraft velocity increases with elevation in this part; i.e. $\frac{dU}{dZ}$ is positive. Four situations can be identified. (a) If $U > \vartheta_r$, where U is the updraft velocity and ϑ_r is the fall velocity of a drop of radius r, drops will move upward. (b) If $U = \vartheta_r$, drops will be balanced. (c) If in addition to $U = \vartheta_r$, $\frac{dU}{dt} = \frac{d\vartheta_r}{dt}$, then the drops will start accumulating. However, this condition is very, very stringent and critical. It does not seem likely that the rate of increase in the updraft velocity will exactly match the rate of increase in fall velocity of the growing drops for significant duration. Thus the conditions (b and c) if at all satisfied will be for a very short duration and hence it is unlikely that high accumulations of hydrometeors would result. (d) If $U < \vartheta_r$ drops will fall out since $\frac{dU}{dZ}$ is positive in this region. Thus, most of the drops in the lower part of the updraft, due to their small sizes, will be moving upwards; some which grow rapidly may fall out. Therefore, the accumulation of millimeter size drops in the lower part does not seem possible.

8.3.2.2 Growth and accumulation in the upper updraft region:

In the upper part of the updraft profile $\frac{dU}{dZ}$ is negative and drops will behave in a fashion similar to the behavior in the lower part for the conditions (a), (b), and (c) previously describe However, when $U < \vartheta_r$, drops will move downward with velocity $(\vartheta_r - U)$ since $\frac{dU}{dZ}$ is negative. Hence the drops will be continuall balanced as they try to fall. Thus, as the drops grow from say 1 m to 5.5 mm diameter, they will keep falling slowly but still be supported by the updraft. As long as the updraft maximum is greater than 12 m sec^{-1}, the fall velocity of the largest stable drop, all drops that are trapped in the upper part of the updraft cannot escape; thus they keep growing and are balanced.

The fall velocities of all drops greater than 4 mm are nearly equal and this fact seems to contribute to the high accumulation of hydrometeors in a region small in vertical extent. Moreover, as the developing precipitation increases the drag on the updraft, the slope $(\frac{dU}{dZ})$ increases, resulting in a large accumulation of liquid water as shown in Fig. 8-6. The hydrometeors in the cloud depth Z_2 are squeezed in a smaller cloud depth Z_1. If one assumes average liquid water content in cloud depth Z_2 then it is possible to calculate its increase when the depth Z_2 decreases to Z_1. Thus depending on the ratio Z_2/Z_1 and the average liquid water content, one can expect accumulation of water contents in excess of 30 to

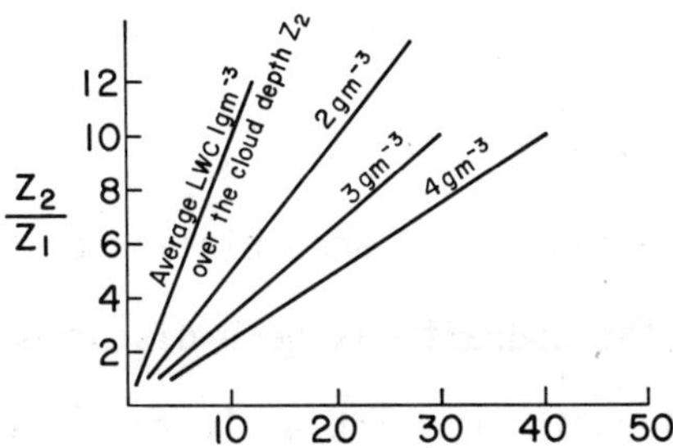

AVERAGE LWC (g m^{-3}) OVER THE CLOUD DEPTH Z_1

FIG. 8-6 Accumulation of high liquid water content in the core. (from Gokhale and Rao, 1969a)

40 g m^{-3} in the core. Of course, such accumulation is localized in time and space and requires intense and persistent updraft (Gokhale and Rao, 1969b). Such high values of LWC have been recently found (NCAR, 1972) as stated in section 8.3.2. For the effect of horizontal divergence on accumulation of particles in the core, refer to discussion in section 8.9.

8.4 Origin of hail embryos:

The mechanism of the accumulation of hydrometeors in the upper part of the updraft has been discussed in the earlier section. An accelerating updraft would take the drops to higher and higher regions of lower temperatures. It may be pertinent her to refer to the model cloud diagram (Fig. 8-2). The lines a, b and c indicate, as discussed in section 8.2.3, freezing probabilit and temperatures of supercooled drops of different sizes. So, depending on the probability of freezing as indicated by the three lines, the drops would freeze between the temperature range -4°C and -30°C. These frozen drops form the embryos of hailstones and grow further by accreting the supercooled cloud droplets carried up in the updraft.

The sizes of the hail embryos range from 100 μ to about 6.0 mm, and the embryos may originate at any height depending on their position in the updraft and the probability of freezing. It is easy to distinguish two regions in the updraft, first the regic where the frozen or supercooled drops accumulate and second, the region where the embryos grow to the hailstone sizes. In Fig. 8-3

ion A indicates where accumulation of both supercooled and
zen drops of diameter up to 6.0 mm takes place. These drops
m the embryos and because of their higher terminal velocities
l out of region A into the lower region B and grow to hailstone
e.

The formation of graupel in region A depends on the
bability of freezing of smaller-size droplets. Such frozen
plets or crystals grow further by accretion of micron
e supercooled droplets as well as of ice crystals and result
graupel particles. Once a graupel particle attains diameter
ater than 6 mm, it will fall to region B and eventually would
w to a large hailstone. Thus, a hail embryo can be either a
upel or a frozen drop.

. Region of hail growth:

The regions of embryo and hail growth in the model
ud (Fig. 8-2) extend in the vertical from approximately 3.5 km to
km with temperature varying between -4°C to -30°C. The regions
further characterized by the updraft decreasing with
reasing height. Thus, there exists an air-mass divergence
this region when the updraft is steady. The horizontal
placement of the hail embryos during their growth due to
s effect is discussed in Sec. 8.9.

As stated earlier hail embryos originate in region A, and
w to larger sizes and fall to a lower level and enter region B (Fig. 8-3)

because of their higher fall velocities than those of millimeter size supercooled drops, which are suspended in region A. Hence hail embryos in region B cannot grow by accreting millimeter size drops. The growth is due to accretion of tiny supercooled droplet carried toward the embryos from the lower part of the cell. (The only exception is the growth of hailstones with icicle lobes as discussed in Sec. 5.9.2). The growth of cloud droplets originating at the base of the cloud is calculated in the next section. These are carried upward in a strong updraft towards region B (Fig. 8-3) where the hail embryos are balanced.

8.6 Growth of cloud droplets in a strong updraft:

The rate of increase of mass of a drop of radius r growing both by condensation and by coalescence with smaller droplets is given by the equation:

$$r \frac{dr}{dt} = G \left(S - \frac{a}{r} + \frac{b}{r^3}\right) f(Re, Pr') + \frac{r E \vartheta w}{4 \rho_L} \qquad (8\text{-}5)$$

where

$$G = \frac{D \rho_V}{\rho_L} \left[1 + \frac{D L^2 \rho_V M_0 f(Re, Pr')}{R T^2 k f(Re, Pr)}\right]^{-1} \qquad (8\text{-}6)$$

S is supersaturation of the environment;

D is coefficient of diffusion of water vapor in air;

ρ_V is density of water vapor in the environment;

ρ_L is density of liquid water;

L is latent heat of vaporization of water;

M_0 is gram molecular weight of water;

f(Re, Pr') and f(Re, Pr) are ventilation factors;

R is the gas constant;

k is thermal conductivity of air;

T is ambient temperature;

a/r is Kelvin curvature factor;

b/r^3 is nucleus solubility factor;

w is concentration of liquid water in the form of small droplets;

E is collection efficiency;

ν is fall velocity of the droplet.

The first term on the right hand side of equation (8-5) is due to Fletcher (1962) and gives the growth rate by condensation. The second term is the well-known coalescence term.

The growth of a cloud droplet is calculated in the model cloud (Fig. 8-2). The supersaturation is assumed to be 0.5 percent and adiabatic liquid water content as shown in Fig. 8-4. First the droplet growth due to condensation alone is calculated until it reaches a size of 40 μ. Then the growth due to condensation as well as coalescence is considered. These calculations are made at one minute intervals starting from the cloud base. The position of the drop in the updraft at the end of each minute is estimated from average updraft velocities in the model cloud. The growth rate between any two levels is calculated by assuming average liquid water content between those two levels. Figure 8-7 shows the curve for droplet diameter versus height. The droplet size increases from 60 μ to 300 μ between the levels 4.5 and 7.5 km. The size of a supercooled droplet reaching an embryo would naturally depend on the location of the hail embryo; however, it would be in the size range of 60 to 300 μ if the initial size is 30 μ.

8.7 Hailstone growth curves:

Hail embryos grow by accreting rising cloud droplets. The growth due to accretion is calculated with the help of the well-known equation:

$$\frac{dr}{dt} = \frac{Ew\,(\vartheta_H - \vartheta)}{4\rho_i} \qquad (8\text{-}7)$$

The values of E, the collection efficiency, and ρ_i, the ice density, are assumed to be 1.0 and 0.9 respectively. w is the liquid water content of the cloud (adiabatic value) and ϑ_H and ϑ represent the fall velocities of hail and water droplets respectively. (Ew) is the effective liquid water content being the product of the collection efficiency and the liquid water content.

Since the embryos are suspended in the upper part of the updraft as shown in Fig. 8-3, there is sufficient time for the embryo to grow to a realistic size. Figure 8-8 illustrates the growth of a hail embryo of 5 mm diameter in 20 minutes for three different values of effective liquid water content (Ew). These calculati

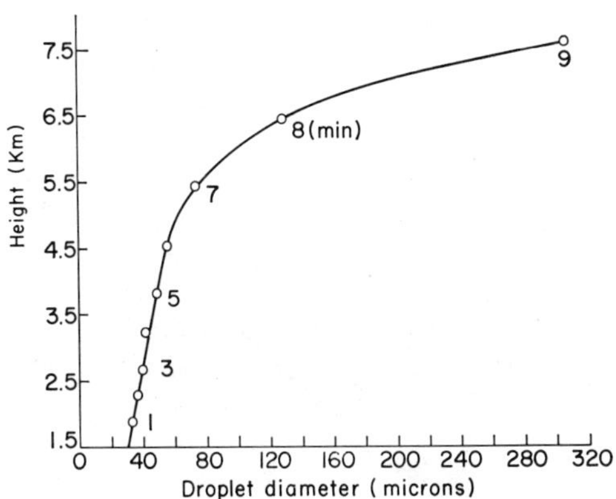

FIG. 8-7 Growth of a cloud droplet whose initial size is 30 μ; supersaturation = 0.5 percent and adiabatic liquid water content for a model cloud with base temperature +10°C, as shown in Fig. 8-4 are assumed. (from Gokhale and Rao, 1969a)

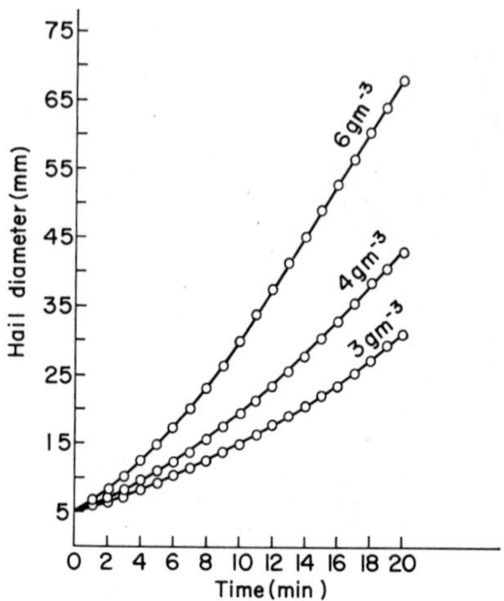

FIG. 8-8 Growth of hail for three different values of effective liquid water content. (from Gokhale and Rao, 1969a)

are carried out at one minute intervals. It can be seen from Fig. 8-4 that LWC is fairly constant in the region where major hail growth takes place as discussed in section 8.2.5. The size of the hailstone after 20 minutes will be about 4.2 cm with an average (Ew) of 4.0 g m^{-3}, E being assumed to be unity. Depending on the cloud base temperature and height, the liquid water content varies between 3 and 6 g m^{-3} and in extreme cases up to 8 g m^{-3}. In Fig. 8-8 hailstone growth curves for (Ew) having values 3 and 6 g m^{-3} are also included. In the latter case, a hailstone will attain size of 6 to 8 cm in 20 minutes.

As the hailstone grows by accretion it will slowly fall in the upper part of the updraft, $(\frac{dU}{dZ})$ being negative, and will continue to grow, being unable to leave the updraft. As mentioned earlier, as long as the updraft velocity maximum is greater than the fall velocity of a hailstone, it cannot leave the updraft which may be either pulsating or quasi-stationary. Thus the hailstone finds sufficient time in a strong and persistent updraft to grow to a realistic size. A pulsating updraft would take it to higher or lower levels. However, if the updraft were steady, then it would remain within a small vertical region. Thus, the nature of the updraft, whether pulsating or steady, does not seem to inhibit in any way the growth of a hailstone, provided, of course, that the pulsation was such that the stone was not able to fall out during a low-velocity period of the updraft.

8.8 The development of internal structure:

The types of ice deposit that can be distinguished in the complex structures of hailstones have been described in Section 5.3. During the hailstone growth process, the latent heat of fusion of the accreted water is liberated. If it raises the temperature of the surface of the stone to 0°C, the growing hailstone develops a wet skin. The two basic growth processes responsible for the development of internal structure of hail are shown in Table 8-1. A hailstone layer growing in a wet and warm environment is expected to have medium to large crystals, whereas growth in a dry and cold environment or with the accretion of ice crystals should result in small crystal fabric.

8.8.1 Dry growth:

Dry growth results from the collection of cloud droplets or drizzle-size drops which freeze individually. Small air bubbles are embedded in the ice making its appearance opaque. Small grains (crystallites) may be formed also by accretion of ice crystals, a process which contributes to random orientation of the c-axis. The crystals are an order of magnitude smaller than those in 'wet growth' ranging from 300 μ in diameter to less than 100 μ. These are ten to thirty times larger than average size cloud droplets.

8.8.2 Wet growth:

Wet growth develops if the heat exchange between the

TABLE 8-1

Two basic growth processes responsible for the development of internal structure of hail.

GROWTH TYPE	SPECIFIC GRAVITY	LAYER APPEARANCE	CRYSTALLITES (grains)	C-AXIS ORIENTATION
dry growth	0.87 or less	opaque	small	random
wet growth (clear or spongy)	0.90-0.92	clear or slightly opaque	medium to large	radial (clear ice) tangential (spongy ice)

growing deposit and its surroundings is not sufficient to freeze all the accreted water. The growth could be either clear or spongy. Ice formed around the embryo is composed of medium to large crystals oriented radially outwards (List 1959a). The crystals may be more than 1 mm in width and 5 mm in length.

List observed with the help of wind tunnel experiments that all the water does not spray off the incompletely frozen hailstone but at least part of it becomes built into the structure of dendritic crystallites which grow inside the accumulated water. He divides the whole range into three subdivisions: porous ice by dry growth, and spongy ice and clear ice by wet growth as shown in Fig. 8-9. Clear ice is indicated by the curves. On the left-hand side of the curves will be spongy growth and to the right of the curves will be dry growth.

How the internal structure develops due to dry and wet growths according to the model proposed by the author is illustrated by the following examples:

8.8.3 Example 1:

Figure 8-10 is constructed to illustrate the basic features of the growth of a hailstone. Let the embryo 5 mm in diameter originate at a level where the cloud air temperature is -12°C. The sequential growth is described below:

A. Embryo (d = 5 mm), temperature = -12°C.

B. Embryo grows in average $W = 4\ g\ m^{-3}$ in a quasi-steady

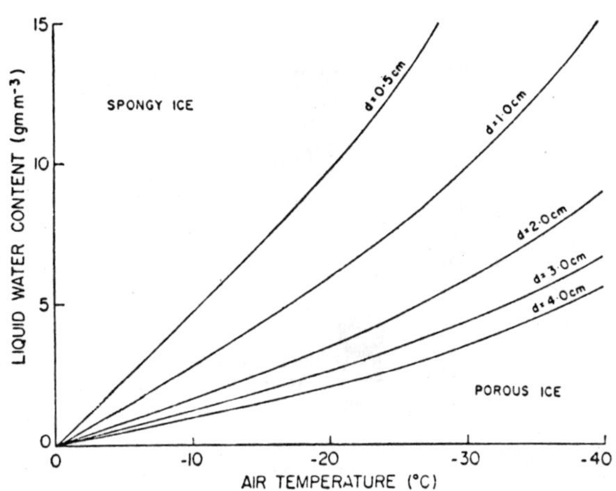

FIG. 8-9 Conditions for the growth of spongy ice, clear ice (corresponding to the curves), and porous ice, in terms of the ambient liquid water content, the temperature and the diameter d of freely falling spherical hailstones, at a pressure of 733 mb. (from List, 1961a)

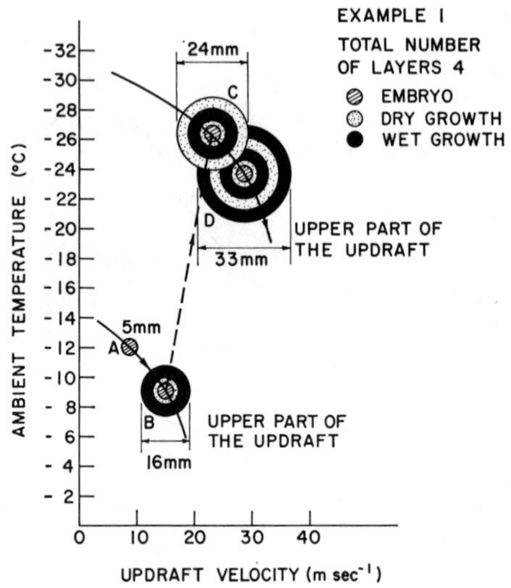

FIG. 8-10 Schematic illustration of wet and dry growth of a hailstone. (from Gokhale and Rao, 1969a)

updraft until it has grown to the size d = 10 mm. Temperature at that level is -9°C. The embryo has sufficient time to grow to this size as it is falling in an increasing updraft.

C. Let us assume that the updraft then develops rapidly and takes the hailstone to approximately -26°C level. During this rise the hailstone continues to grow.

D. The same growth cycle is repeated as described in (B). It results in a size d = 33 mm, and if the updraft decreases at this time then the hailstone will fall out of it. The hailstone will also fall out of it if it has grown to such a size that its fall velocity is greater than that of the updraft.

As discussed earlier, there is little mixing in the core of a well-developed updraft and, therefore, the liquid water content in such a core would have moist adiabatic values. Its variation with height for the model cloud is shown in Fig. 8-4. Thus, assuming 4 g m^{-3} as the average effective liquid water content in the region between -4°C and -30°C where hail is growing, and using List's experimental data as shown in Fig. 8-9, the curve (b) shown in Fig. 8-11a has been constructed by plotting ambient temperature versus hailstone diameter for which the surface temperature of the hailstone would be 0°C. Curve (a) is the growth curve for (Ew) equal to 4 g m^{-3} similar to the one

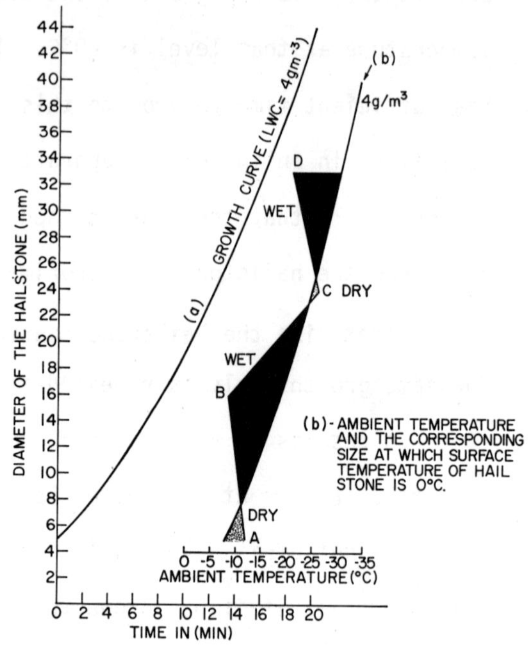

FIG. 8-11a Wet and dry growth of a hailstone; number of layers = 4. (from Gokhale and Rao, 1969a)

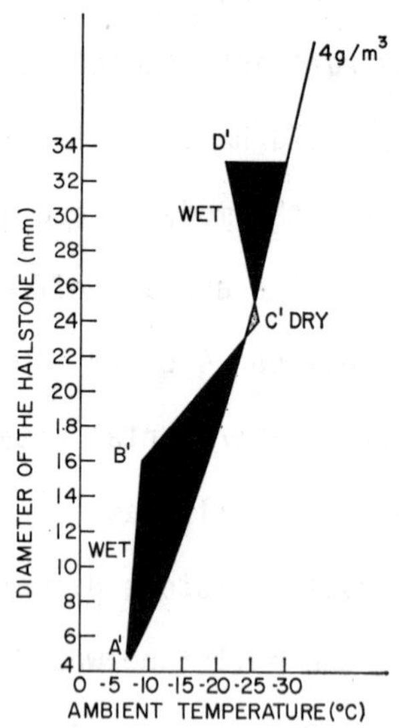

FIG. 8-11b Wet and dry growth of a hailstone; number of layers = 3. (from Gokhale and Rao, 1969a)

shown in Fig. 8-8. The points A, B, C, D representing ambient temperatures versus embryo size are plotted in Fig. 8-11a. The part of the curve A B C D which lies to the right of 'wet and dry' curve (b) indicates dry growth; the part on the left, shows wet growth. Thus, the successive layers are due to alternate wet and dry growths; the outer layer is due to wet growth. The total number of layers is four.

It is interesting to note that the first layer of dry growth and part of the second layer due to wet growth are developed in the first phase of the quasi-steady updraft; the remaining part of the second layer due to wet growth and part of the third layer due to dry growth are developed while the hailstone is being taken up to a higher level, and the remaining part of the third layer due to dry growth and the outermost layer of wet growth are developed during the increased updraft phase.

Therefore, when the growth starts at a comparatively low temperature, two layers, a dry and a wet, develop sequentially as the hailstone is growing and slowly falling against a quasi-steady updraft. The cross section of the resulting hailstone would be similar to the one of natural hail shown in Fig. 5-13(a).

8.8.4 Example 2:

In this case let us assume the embryo A' originated at a comparatively warmer temperature, such as -7°C. It has gone through stages B', C' and D', as shown in Fig. 8-11b. The curve

A' B' C' D' is plotted, as in the first example, and shows how the layers develop. A layer of wet growth is followed by a dry growth layer; the outer layer is due to wet growth. The total num of layers resulting is three. The cross section of the resulting hailstone would be similar to the one of natural hail shown in Fig. 5-13(d).

8.8.5 Layer growth in general:

These examples show that only one substantial rapid increase in updraft velocity is required to develop 3 or 4 layers. Most of the hailstones found in nature show 3 or 4 layers; but 10 or more layers are sometimes detected. Such a large number of rather thin layers would develop due to a number of rapid increases and decreases in the updraft velocity, resulting in the ambient temperature criss-crossing the 'wet and dry' curve.

Another possibility has been suggested in the literature for the development of internal structure, namely, the variation in liquid water content. But as discussed in section 8.2.5 the variation in LWC in the region where the major hailstone growth takes place is rather small. On the other hand, variation in ambient temperature is quite significant. Moreover, Levi and Aufdermaur (1970) suggest (as discussed in section 5.6) that essentially <u>only changes in air temperature can cause large variations in crystal size, which result in a layered structure</u>.

We suggest therefore that multiple fluctuations of the updraft and the resulting changes in ambient temperature of the hailstone are mainly responsible for its multi-layered structure.

If the ambient temperature remains to the right of the 0°C curve throughout hailstone growth, then opaque ice due to dry growth would result. If it remains to the left during the entire growth, then only wet growth would appear. This is how some hailstones develop without having alternate layers of wet and dry growth. Besides, structures are found in natural hailstones (though rather rare) with large protrusions in butterfly shape as shown in Fig. 5-7. This growth may occur at slightly higher water contents and at comparatively colder ambient temperatures, as discussed in Sec. 5.9.2.

8.8.6 Layer of wet growth and evaporation while hailstone descends:

When the fall velocity of a hailstone becomes greater than the updraft velocity maximum, the hailstone descends toward the 0°C isotherm. The outer layer at the start of the descent may be wet or dry. However, while descending, it would have an outermost layer due to wet growth, the thickness of which would depend upon the size of hailstone. Below the base of the cloud, this layer would start melting and evaporating as the hailstone falls into a warmer environment. The outer layer, if thin, might melt completely and hence not be detectable in a hailstone picked up on the ground. For the discussion on the melting of hailstones, see Appendix B.

8.9 The horizontal divergence:

Let us now discuss the effect of horizontal divergence on particle growth and accumulation.

The horizontal divergence of an air mass is related to the vertical velocity by the equation of continuity. Hailstorms have unusual vertical motion, and the values of divergence are higher than 10^{-4} sec^{-1} (Weickmann, 1969).

From continuity equation it follows that:

$$\frac{\partial v}{\partial x} + \frac{\partial w}{\partial y} = - u \left(\frac{1}{\rho} \frac{\partial \rho}{\partial z} + \frac{1}{u} \frac{\partial u}{\partial z} \right) \qquad (8\text{-}8)$$

where v and w are the components of wind velocity in x and y direction; ρ is the density of the air and z is the height; u is the vertical velocity.

Assuming a radially symmetrical jet, the horizontal velocity V_S at a distance S from the axis can be written from equation (8-8) as:

$$V_S = - \frac{S}{2} \left(\frac{u \partial \ln \rho}{\partial z} + \frac{\partial u}{\partial z} \right) \qquad (8\text{-}9)$$

$(\partial \rho / \rho \partial z)$ is always negative and has a value of -1.25×10^{-4} in the MKS system. Haman (1967) and Phillips (1969) have calculated the horizontal displacement of particles using equation (8-9) and assuming $(\partial u / \partial z)$ constant in the upper part of the updraft.

8.9.1 Radial velocity (V_S):

The equation (8-9) gives the radial velocity of the

hydrometeor since only the particle inertia acts against the air drag force in the horizontal direction as discussed by Phillips (1969). Since the velocity is a linear function of S, it is evident that if the updraft is approximately vertical, hydrometeors on the axis remain on the axis. For higher values of S, for a particle suspended in a steady updraft, V_S can be calculated with equation (8-9). But the updraft in a developing thunderstorm cell is pulsating and accelerating before it attains a steady or quasi-steady state as discussed in Sec. 8.1.

8.9.1.1 Pulsating updraft:

For increasing and decreasing updraft velocity gradients of equal magnitude, the value of V_S depends only on the density term, the first term in the bracket on the right-hand side of equation (8-9); its magnitude is of the order of 10^{-4} to 10^{-3} sec^{-1}. Thus, there would be net horizontal displacement of hydrometeors in a pulsating updraft.

8.9.1.2 Accelerating updraft:

When the updraft is accelerating, $(\partial u/\partial z)$ term is positive whereas the density term, as pointed out earlier, is negative. The order of magnitude of $(\partial u/\partial z)$ term is 10^{-3} to 10^{-2} sec^{-1}. Thus, if $(\partial u/\partial z)$ is greater than $(u\partial \ln\rho/\partial z)$, the hydrometeors experience convergence. If the two terms are equal in magnitude, then the net displacement would be zero.

8.9.1.3 Steady updraft:

For a steady updraft the value of $(\partial u/\partial z)$ in the upper part is negative. Thus both the terms in the bracket on the right-hand side of equation (8-9) are negative, and hence, V_S has a positive value. In other words, the hydrometeors would always experience horizontal divergence in the upper part of a steady updraft.

For a steady updraft shown in Fig. 8-3, the value of $(\partial u/\partial z)$ in the upper part is not constant but increases from 10^{-4} sec^{-1} near the height of maximum updraft to 10^{-2} sec^{-1} at upper level. Figure 8-12 shows the radial velocities (V_S) of hydrometeors at different heights (Z) for three different values of S calculated with the help of equation (8-9).

8.9.2 The displacement of particles in a steady updraft with a specific example:

Making use of Fig. 8-12, one can estimate the maximum distance from the axis that a hailstone can travel and still stay within the core. Beyond this limit a hailstone will leave the core in a certain interval of time. Consider as an example, a 24 mm diameter hailstone in a steady updraft profile as illustrated in Fig. 8-10 (upper half) in a core of radius 500 m and having liquid water content of 4 g m^{-3}. This hailstone takes approximately 4 minutes to grow to 33 mm and falls to the level of maximum updraft. The horizo

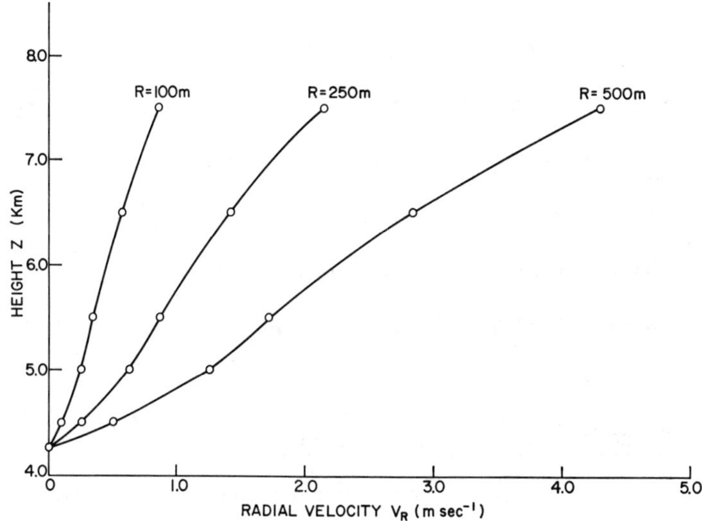

FIG. 8-12 Radial velocities at different heights for three values of S, the distance from the axis of the core. (from Gokhale and Rao, 1969a)

displacement suffered by this hailstone during this time is calculated at one minute intervals considering its height and radial distance from the axis of the core at the end of each minute. A hailstone that was located originally at a distance of 200 m from the axis of the core suffers a displacement of about 650 m during its 4 minute fall, whereas a hailstone located at 160 m suffers a displacement of about 500 m during this time interval and reaches just the periphery of the core. Thus, the hailstones that were originally located up to 160 m from the axis will not leave the core.

8.9.3 Concentration of large hailstones:

The number of hail embryos remaining in the core per unit volume is given by

$$N' = N'_0 \, (S_2/S_1)^2 \qquad (8\text{-}10)$$

where S_1 is the radius of the core of the updraft;

S_2 is the maximum distance of original location of the hailstone from the axis, so that it does not leave the core;

N'_0 is the concentration of hail embryos per unit volume in the core, when the updraft becomes steady.

The embryos (N') will grow to large sizes, overcome the updraft velocity maximum, and fall to the ground. So a rough estimation of this number can be made. Thus, with S_1 = 500 m,

S_2 = 160 m (the value estimated in the earlier section) and N'_o = 1 to 10 m^{-3} (Blanchard and Spencer, 1970), then number (N') of large hailstones will be between 0.1 and 1.0 per m^3. Concentrations of this magnitude have been found (see Sec. 2.8).

8.10 Wind shear and the growth of hail:

Based on observations of cumulus clouds and radar echoes which were tilted downwind due to wind shear, Byers and Braham (1949) concluded that vertical wind shear was harmful to storm development. On the other hand, individual hailstorms with strong vertical shear at upper levels have been documented by Ludlam and Browning (1960) and Fujita and Byers (1960). Dessens (1960) has suggested (see Sec. 7.2.6) the presence of a very strong wind between 6 and 12 km as the factor which determines whether or not a thunderstorm will transform itself into a destructive hailstorm. It is probable that a strong vertical wind shear in the upper troposphere may actually invigorate a thunderstorm and prolong its life (Newton, 1963; Fankhauser, 1971). It is observed that severe storms embedded in strong shear often persist for many hours even at night when air-mass type thunderstorms tend to die out.

However, time-lapse movies reveal that as long as a cloud rises vigorously, it remains upright and moves in an upright position with the wind field, but that it will be bent over in the wind field after the updraft slows down (Weickmann, 1964). Furthermore, Hitschfeld (1960) called attention to the radar observation that hailstorms display a remarkable vertical stability and remain upright even in a strong wind shear.

The analysis of the relative occurrences of thunderstorms and hailstorms in relation to vertical wind profile has been reported by Ratner (1961) who concludes that in the United States "neither speed of the winds aloft nor the wind shear between 500 and 200 mb appear to be determining factors in occurrences of hail."

The proposed model shows that the growth and internal structure of hailstones can be satisfactorily explained without invoking the presence of wind shear between 500 and 200 mb levels. If the wind shear is pronounced in the region of the cloud where hail growth is taking place, then it is likely that hailstones will be thrown out of the tilted updraft and hence their growth to large sizes will be hindered. This has been confirmed by the observations reported by Sulakvelidze (1967).

8.11 Conclusions:

The new and essential features of our model are summarized below, and they indicate how this model differs from earlier models of hailstone growth discussed in Chapter 7.

The embryos of hailstones originate at upper levels in the region of high accumulation of liquid water content. In this respect the model differs from the one proposed by Ludlam (1961), where he assumed the origin of embryos near 0°C isotherm.

The growth of hailstone embryos, because of their higher fall velocities compared to those of rain-size supercooled drops, takes place at lower levels (though still above the updraft maximum) and thus they do not grow in high liquid water content. Their growth from about 6 mm on is at the expense of supercooled water droplets sweeping by in the updraft. These embryos are suspended in the upper part of a strong updraft and, as they continue to grow, they fall slowly against an increasing updraft. This is how embryos get sufficient time to grow to large sizes. Thus the elaborate recycling process in a tilted updraft due to wind shear, suggested by Ludlam (1961), is not essential for hail growth.

The growth of embryos takes place in LWC of 3 to 6 $g\ m^{-3}$ and not in very high LWC as proposed by Hitschfeld and Douglas (1963) and by Sulakvelidze (1967). LWC of 3 to 6 $g\ m^{-3}$ is adequate to obtain large hailstones in about 20 minutes, starting with an embryo size of 5 mm.

Hailstone growth takes place mainly at levels where the temperatures are between -4°C and -30°C. Above the -30°C isotherm, most drizzle-size supercooled drops freeze and hence the hail growth substantially decreases.

Our model also accounts for the internal structure of a hailstone, namely the presence of alternate layers due to dry and wet growths. It postulates that multiple fluctuations of the updraft and resulting changes in ambient temperature of the hailstone are mainly responsible for its multi-layered structure.

It has been shown that the effect of horizontal divergence on particle growth accounts for the low concentrations of hailstones as compared to those of millimeter size embryos.

The proposed model is one dimensional and the profiles and velocities of updrafts are assumed, which are confirmed by the field measurements. But as already pointed out in section 8.1, the medium and large size hailstone growth takes place in the core of the cell, which is characterized by little mixing (Ludlam, 1961; and Davies-Jones, 1974). Thus it is reasonable to assume that the parameters such as temperature and LWC are constant across the core of the cell. Thus the growth of medium and large hail can be accounted for with the help of a one-dimensional model, taking into account the variation of different parameters only in the vertical.

The variation in LWC (adiabatic) in the region, 4 km to 8 km, in the core of the cell where the major hailstone growth takes place, as discussed in section 8.5, is rather small as seen from Fig. 8-4; hence, in this region, the average value of LWC can be assumed for the calculations of hailstone growth. It is not the small variation of LWC but the temperature variations of the ambient environment due to multiple fluctuations of the updraft that are mainly responsible for the multi-layered structure of hailstones.

CHAPTER 9

ISOTOPIC ANALYSIS OF HAILSTONES

As discussed in Chapter 5, the interpretation of the bubble and crystalline structures of hailstones depends upon a number of variables, such as opacity of ice formed under various conditions, droplet temperature, droplet size and concentration, speed of impact, and wet and dry growth regimes. Therefore, only broad interpretations of the bubble and crystalline structures are possible. On the other hand, the isotopic method of analysis has the advantage that the ambient temperature is virtually the only variable involved in the interpretation of the data (Macklin et al., 1970). From such an analysis the air temperatures, updraft speeds and trajectory of a hailstone in the cloud can be inferred.

9.1 The isotopic analysis:

The isotopic technique to study hail formation by the determination of deuterium content in hailstones of various dimensions was first developed by Facy, Merlivat, Nief and Roth (1963) and later applied by Merlivat et al. (1965) and Majzoub et al. (1968). Recently it has been used by Macklin, Merlivat and Stevenson (1970), to determine the structure and life history of a single large hailstone which grew in a strong updraft of a severe storm.

As discussed in earlier chapters, natural hailstones may possess alternate layers of clear and opaque ice. Most of the frozen layers remain unmixed once formed, except for the outer layer, part of which can melt during the fall in the subcloud region.

The deuterium content of the liquid or solid particles formed in the cloud is directly related to the temperature at which condensation takes place. If a model of the cloud is assumed then it is possible to calculate the altitude at which this process occurs.

9.2 Theory:

Each molecular species has a different equilibrium vapor pressure (Friedman et al., 1964). As condensation of water vapor occurs within a cloud, the liquid phase acquires a greater concentration of HDO molecules than the vapor phase. The relative enrichment in droplets is expressed by a separation factor α, where α is the ratio of the mole fractions of HDO molecules in the liquid and vapor phases respectively. As shown by Facy et al. (1963), the enrichment of cloud droplets decreases with height for two reasons. First, the ratio of the mass of liquid to mass of vapor increases with height; second, α is temperature dependent.

To facilitate the interpretation of the isotopic composit Macklin et al. (1970) made the following assumptions:

(i) The hailstone grows in air making a saturated adiabatic ascent and therefore the molecular species $H_2{}^{16}O$, $HD^{16}O$ and $H_2{}^{18}O$ are conserved in the air.

(ii) The relations between temperature, pressure and height are specified by a wet adiabat.

(iii) The relaxation times for the exchange processes of cloud droplets are fractions of a second. Therefore the ascending cloud droplets are in isotopic equilibrium with the vapor at all temperatures.

Under these assumptions, the isotopic concentration of the liquid can be expressed as

$$\delta D = \frac{N_L}{N_{ST}} - 1 \qquad (9\text{-}1)$$

where N_L and N_{ST} are the mole fractions of HDO molecules in the sample and Standard Mean Ocean Water, respectively. Following the treatment given by Macklin et al. (1970), the value of δD can be written as

$$\delta D = \frac{\delta D_v^i + \left(\frac{\alpha - 1}{\alpha}\right)\frac{m_v}{m_o}}{1 - \left(\frac{\alpha - 1}{\alpha}\right)\frac{m_v}{m_o}} \qquad (9\text{-}2)$$

where δD_v^i is the initial deuterium value for the vapor. α is the ratio of the mole fractions of HDO molecules in the liquid and vapor phases, m_v and m_o are the residual mass of vapor and the total mass of water substance, respectively.

Merlivat and Nief (1967) have measured the values of

α for HDO for temperatures down to -20°C when vapor was in equilibrium with liquid. These values can be extrapolated reliably to -40°C. This it is possible to calculate from equation (9-2), the values of δD_L as a function of height with knowledge of δD_V^i at a specific temperature in the cloud and the appropriate wet adiabat (Merlivat et al., 1965).

9.3 Experimental method:

9.3.1 The mass spectrometer:

An automatic mass spectrometer for the isotropic analysis of hydrogen in water samples has been described by Lohez et al. (1970). This mass spectrometer which has been further refined and automated by Merlivat can analyze 32 samples of water in less than 24 hours without any assistance from an operator. The instrument is capable of measuring D/H ratios to a relative accuracy of 0.5°/°° (part per thousand) for samples with concentrations of the order of 150 deuterium atoms per million of hydrogen atoms. This is undoubtedly greater precision than that possible with a classical, non-automatic instrument.

The techniques of introducing the samples into the mass spectrometer consists of using a microsyringe enabling analysis of samples as small as tens of mm^3. The possibility of programming a number of samples for the analysis and the number of measurements to be made on each makes the instrument highly flexible.

9.3.2 The experimental procedure:

The hailstone selected for analysis is shown in Fig. 9-1. Its approximately spherical symmetry simplified the analysis and subsequent calculations. It was cut into two sections. One section was photographed under polarized light to show the crystal structure of clear and opaque layers, and their diameters, as seen from Fig. 9-1. The deuterium composition of the other section was measured by the technique described by Merlivat et al. (1965). Small sections 2 to 3 mm in width were cut with a saw in a cold room. The spacing of the cuts was varied in order to improve resolution in the data. Samples from each of the layers were placed in stoppered pyrex tubes and allowed to melt. The direct injection of the water into the mass spectrometer gave the deuterium analysis.

9.4 Analysis of isotopic data:

The results, plotted as δD as a function of hailstone radius, are shown in Fig. 9-2. The δD values are given in parts per thousand and the standard error is $\pm$ 0.7 ‰. The corresponding internal structure of alternate clear and opaque rings was available from optical observations.

The interpretation of the isotopic data was possible when adiabatic ascent was assumed. In the core of the updraft, where little mixing takes place, this assumption is reasonable as discussed by Ludlam (1963).

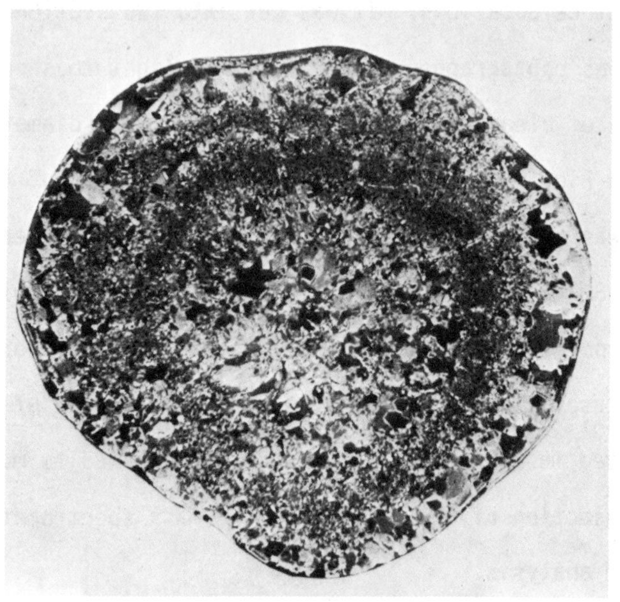

FIG. 9-1 Crystal structure of the hailstone section. Magnification: X 1.4. (from Macklin, Merlivat and Stevenson, 1970)

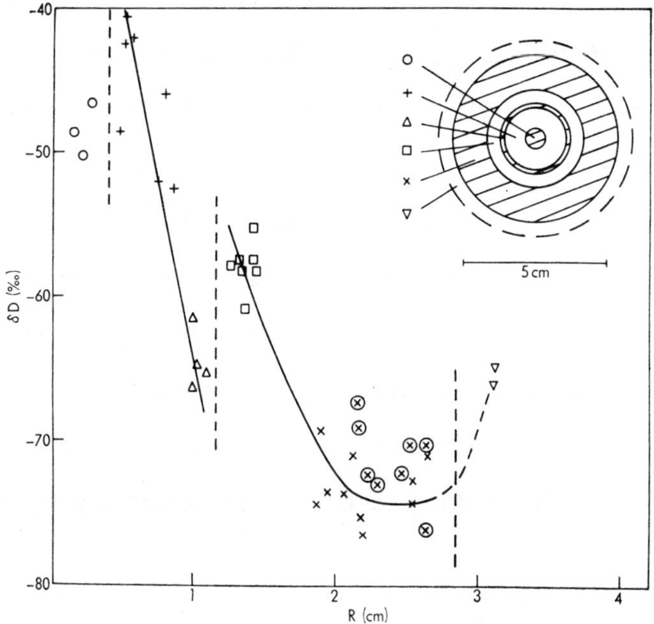

FIG. 9-2 The measured values of δD in parts per thousand as a function of hailstone radius. The equivalent spherical model of the stone is depicted in the upper right-hand corner and the symbols used for samples from the various layers are as indicated. The opaque layers are hatched. The encircled crosses are the δD values for samples comprised in part of the clear ice layer lying within the thick outer opaque layer. (from Macklin, Merlivat and Stevenson, 1970)

It was essential to construct curves of δD versus height to interpret the isotopic data. A method similar to that used by Merlivat et al. (1965) was adopted. The four curves constructed are shown in Fig. 9-3. The lowest curve is based on the assumption that the highest δD measured was produced at 0°C. The other three curves are based on the assumption that the lowest δD measured was produced at -30, -35, and -40°C, respectively. With the help of data illustrated in Fig. 9-2 and 9-3, the height of the hailstone (or temperature at that altitude) was then plotted against its radius, which is the growth trajectory of the hailstone (Fig. 9-4).

9.4.1 The growth trajectory of a hailstone:

It is noted by Macklin et al. (1970) that the growth trajectory shows two main growth zones (marked 1 and 2 in Fig. 9-4) each of which gave rise to two growth layers in the stone. These are alternately clear and opaque with accompanying changes in the crystal size from large to small. The insufficient resolution in the isotopic data would not allow them to explain how the core grew to its full size (7 mm diameter) by an initial ascent. After the first ascent the stone is thought to be carried forward from the top of the updraft enabling it to reenter the cloud at a lower level. It has been suggested that this is essentially the recycling process in the Browning-Ludlam (1962) model of a severe storm. The second ascent begins at a higher level than the first. The flattening of the second part of the curve is due to the

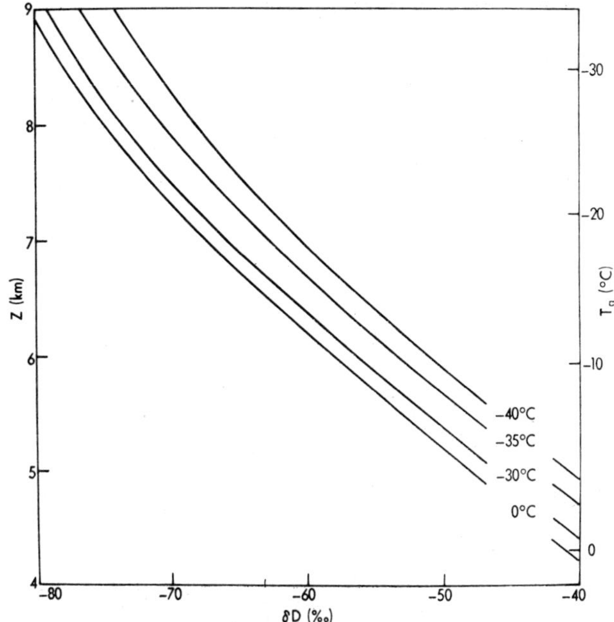

FIG. 9-3 Calculated δD versus height curves. The lowest curve is based on the assumption that the highest δD measured (-40 $^{o}/oo$) was produced at 0°C. The other three curves are based on the assumption that the lowest δD measured (-76 $^{o}/oo$) was produced at -30, -35, and -40°C respectively. (from Macklin, Merlivat and Stevenson, 1970)

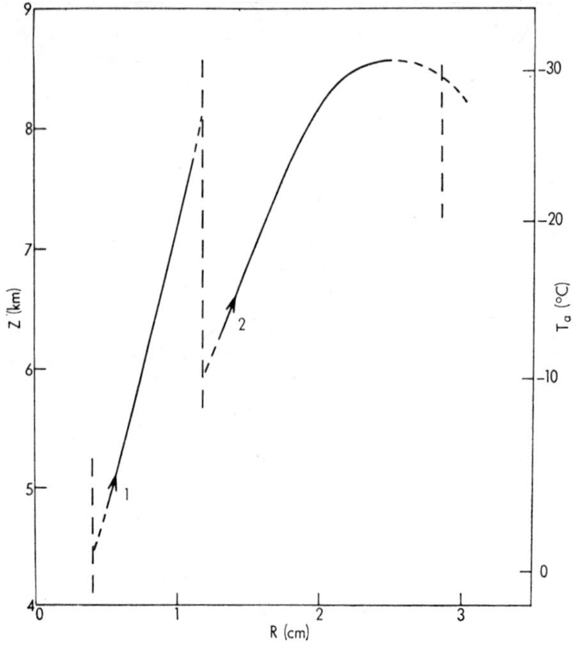

FIG. 9-4 The height of the hailstone as a function of its radius showing the two ascents. (from Macklin, Merlivat and Stevenson, 1970)

fallspeed of the stone becoming comparable in magnitude with the updraft.

9.4.2 Comments on 'The Analysis of a Hailstone':

As discussed in the earlier sections, Macklin, Merlivat and Stevenson (1970) have successfully used the isotopic technique to determine the air temperatures, updraft speeds and liquid water concentrations experienced by a stone during its growth. Moreover they correctly point out that the interpretation of the data of the isotopic method is simpler and more accurate than that of other techniques such as the analysis of the bubble and crystalline structures.

Using the curves of Figs. 9-2 and 9-3, Macklin et al. plotted the variation of the height of the hailstone with radius (Fig. 9-4). The conclusion was drawn that there were two main growth zones numbered 1 and 2 in Fig. 9-4. It is not quite clear why the height corresponding to a $\delta D \approx -50°/_{oo}$ and R $\approx$ 0.27 cm as shown in Fig. 9-2, has not been plotted in Fig. 9-4. Gokhale and Rao (1971) have plotted this point and drawn the curves; they offer a significantly different interpretation of the growth trajectories from that of Macklin et al.

The model suggested by the author for hailstone growth, discussed in Chapter 8, proposes that major growth of embryos takes place in the upper part of the updraft. Further, it has been suggested that the hailstones, do not leave the core of the updraft during their lifetimes; the embryos are locked into the upper part due to the profile of the updraft (namely $\frac{du}{dz}$ is negative in this upper region), and thus the particles get sufficient time to grow to large sizes. They fall out only when their terminal velocities exceed the maximum updraft velocity. (Refer to discussion in section 8.3.2.2.)

With this mechanism of hail growth as a basis and the observed fact that the updrafts in a developing cumulus are pulsating and accelerating, it is possible to construct the trajectory of the hailstone studied by Macklin et al. (1970) as shown in Fig. 9-5. It is suggested that the growth of the hailstone resulted in five successive stages:

<u>Stage I</u>: Embryo (d = 0.54 cm) at -8°C grew to 1.2 cm diameter while falling slowly against an increasing updraft to -5°C. (For the upper part of the updraft $\frac{du}{dz}$ is negative)

<u>Stage II</u>: The stone was then carried upward by the accelerating updraft; it grew to 2.4 cm and reached the height where the temperature was -25°C.

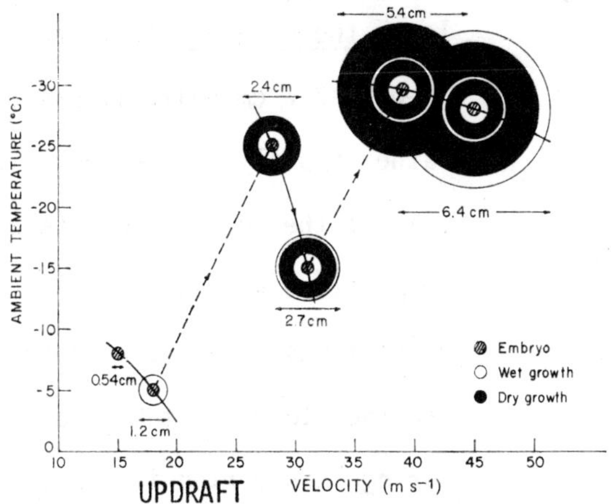

FIG. 9-5 The growth trajectory and development of internal structure of hailstone in the upper part of the pulsating and accelerating updraft. The fall speeds of the hailstone during its growth are taken from Macklin et al. (1970). The updraft velocities which are equal to the fall speeds are plotted on the x-axis. (from Gokhale and Rao, 1971)

<u>Stage III</u>: The hailstone (d = 2.4 cm) <u>did not leave the core of the updraft</u> but grew to 2.7 cm while it was falling rather rapidly in the updraft and reached the lower level where the temperature was -15°C.

<u>Stage IV</u>: Updraft was accelerated once again and the stone was taken to -30°C. During this period it continued to grow and reached a size of 5.4 cm.

<u>Stage V</u>: This is similar to Stage I except hailstone was balanced for some time at one level, then started to fall slowly and grew to 6.4 cm in diameter.

The growths in stages II and IV are similar to the growth suggested by Macklin et al. in ascents 1 and 2 in Fig. 9-4. However, they do not adequately explain the growth in stages I and V. In addition, the growth in stage III was suggested to have occurred while the stone was falling from 7.5 km to 6.0 km level outside the main body of the updraft and was picked up at the lower level where the temperature was -15°C. This explanation by Macklin et al. raises certain discrepancies in the nature of the hailstone growth as outlined below.

(i) According to their analysis, the stone leaves the updraft at a level where temperature is -25°C and is picked up at -15°C level; the distance of fall would be approximately 1.5 km. Assuming its relative fallspeed at 15 m/sec (which would depend on how far away it is from the axis of the cell) the duration of fall would be 100 seconds. Even assuming average value of effective LWC equal to 1 g m^{-3} in the region of its fall outside the main body of the updraft, a hailstone of 2.4 cm would grow to less than 2.5 cm in diameter in 100 seconds. However, as shown by the isotopic analysis, it did grow to 2.7 cm during this descent, a growth rate which would require much higher liquid water content. Hence, this discrepancy indicates that the hailstone did not leave the core of the updraft.

(ii) An updraft profile with a very sharp curvature will be necessary to explain a 2.4 cm hailstone being thrown out of the main part of the updraft at 7.5 km level and then being picked up at the 6.0 km level. It is difficult to justify such a sharp curvature as the relative wind speed values are 11.0 and 9.0 m/sec respectively at those levels as seen from Fig. 9-6.

(iii) Finally, to explain the hailstone growth in stages I and V with a recycling process outside the main updraft would pose similar problems.

Gokhale and Rao (1971) therefore concluded that the hailstone did not undergo the Browning and Ludlam (1962) recycling process as suggested by Macklin et al. to grow to such a large

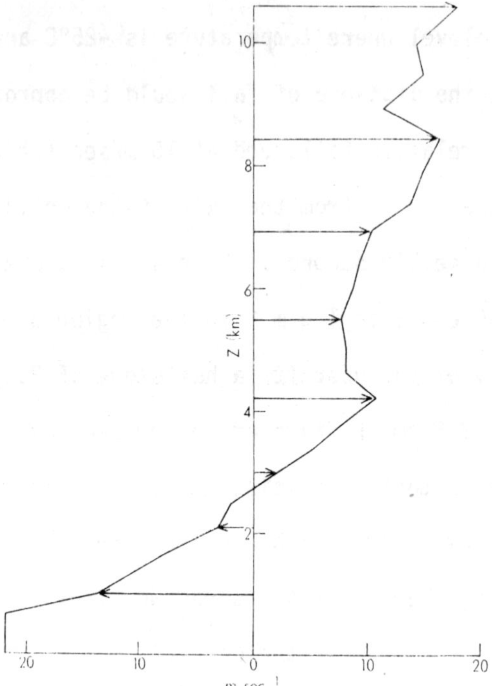

FIG. 9-6 Wind speed relative to the storm. (from Macklin, Merlivat and Stevenson, 1970)

size, but that the entire growth is more likely to have occurred as shown in Fig. 9-5 in the core of the updraft.

Figure 9-5 also shows the development of the internal structure of alternate layers of dry and wet growth which agrees with Fig. 9-1 and the mechanism proposed by the author. (Refer to section 8.8.)

Recently, Jouzel, Merlivat and Roth (1975) have made deuterium and tritium measurements on three hailstones produced during a storm on August 7, 1971 in the province of Alberta, Canada (Fig. 9-7). The variation in the mean altitude formation of each layer of the hailstones A, B, C as a function of their radius is obtained using the deuterium and tritium results. Figs. 9-8, 9-9 and 9-10 show the trajectories of these hailstones during their growth, at least two of the latter occurring in the updraft core, which proves the accuracy of the hail model proposed by the author (Gokhale and Rao, 1971). These data also confirm that the hailstones grew at levels where the temperatures were between -4°C and -30°C.

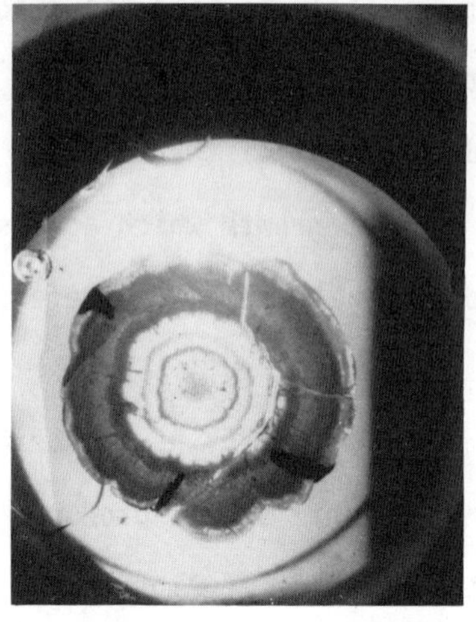

ALBERTA A

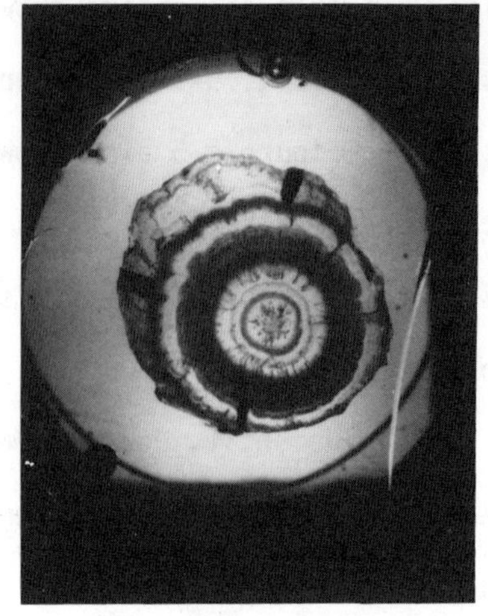

ALBERTA C

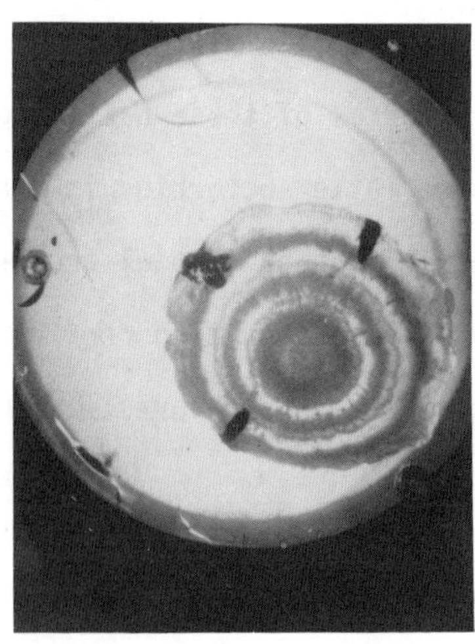

ALBERTA B

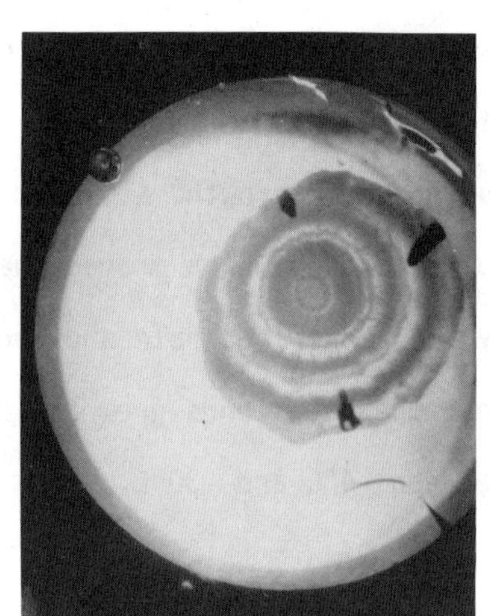

ALBERTA B

FIG. 9-7 Thin sections of three hailstones. (Jouzel et al., 1975)

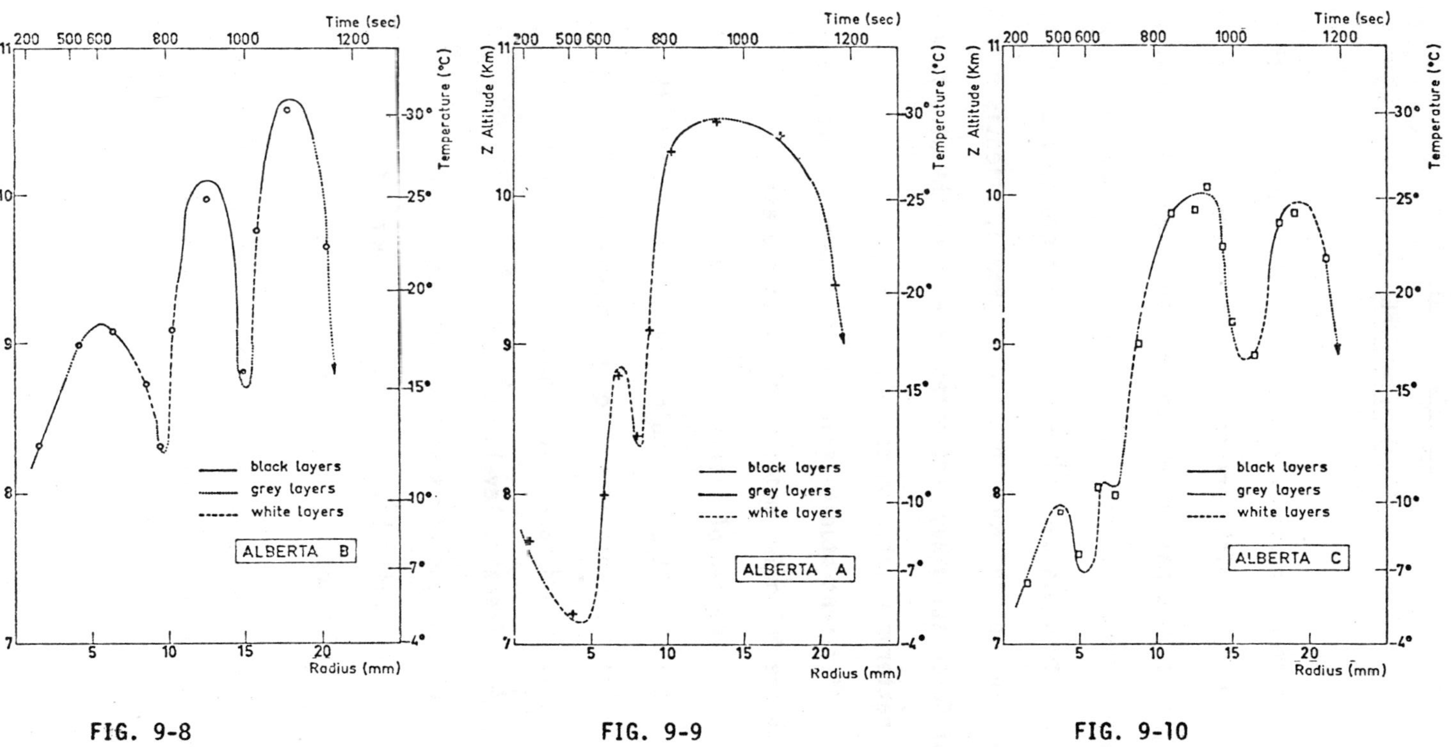

FIG. 9-8

FIG. 9-9

FIG. 9-10

Trajectories of hailstones ALBERTA A; B; and C versus their radii (Jouzel et al., 1975)

CHAPTER 10

HAIL FORECASTING

Meteorologists have recognized for a long time that hail is associated with more severe thunderstorms and that embryos are suspended in a hail cloud long enough to grow to larger sizes because of strong updrafts. Though much of the work discussed in earlier chapters has been directed to the microstructure of the hailstorm and to the processes of hail growth, scientists in the field of forecasting have attempted to analyze the structure of weather patterns which produce hail.

Different studies tried to correlate the occurrence of hail with the following meteorological variables:

(a) Degree of latent instability
(b) Height of wet-bulb freezing level above the terrain
(c) Moisture in the lower atmosphere
(d) Cloud base temperature
(e) Freezing level
(f) Strong winds at high altitudes
(g) Maximum height of radar reflectivity
(h) Intensity of radar reflectivity

The association of hail with one or more of these parameters has been demonstrated but the studies have failed in most cases to reveal the complete picture. It is worthwhile, however, to review the correlations shown by different studies.

10.1 The degree of latent instability:

The simple stability index (Showalter, 1953), which is computed at all radiosonde stations in the USA, is obtained by lifting a parcel from 850 mb level dry adiabatically to its lifting condensation level (LCL), then lifting it saturated adiabatically to the 500 mb level. The resulting temperature is then subtracted from the observed 500 mb temperature. This temperature difference is referred to as the stability index. In general, the stability increases with the increasing value of the index. Some empirical rules used in forecasting are as follows: index 3 to 1--showers probable, thunderstorms possible; 1 to -2--thunderstorms probable; -3 to -5--severe thunderstorms; -6 or less--tornado conditions.

Since hail often results from severe thunderstorms with strong updrafts, a correlation between the occurrence of hail and the degree of instability would seem probable. Fawbush and Miller (1953) and Foster and Bates (1956) used the measure of this instability as a clue to the maximum size of the hailstones. Fawbush and Miller selected 274 upper-air soundings in the United States as representative of air patterns in which hailstones of known sizes are formed. These soundings are grouped according to their severity. The sizes of hailstones and the corresponding updrafts which would support them were estimated from the values calculated by Grimminger (1933) for cases having very little turbulence.

Fawbush and Miller's method relates the size of hailstones found on the ground to the hydrostatic instability of the air mass within which hail-producing thunderstorms occur. Based on the assumption that hailstone growth takes place in the region where the temperatures are from 0°C to -10°C, the positive area (Fig. 10-1) is terminated at the level where the ambient temperature is -5°C. The method of obtaining the curves in Fig. 10-2 for translating upper-air soundings into forecasts of hail size from Fig. 10-2 is described by Fawbush and Miller as follows:

> "The positive area so limited is essentially a triangle, its area is determined by its base and altitude and these are the coordinates of the graph. For convenience, both base and altitude are measured in units of temperature on an adiabatic diagram as illustrated in Fig. 10-1. Assume this figure is the sounding to be evaluated. Determine the mean mixing ratio of the moist layer and follow this saturation mixing ratio curve to its intersection with the temperature curve to fix the convective condensation level (CCL). From the point of intersection, trace the saturation adiabat upward to form one side of the positive area. The temperature curve forms a second side

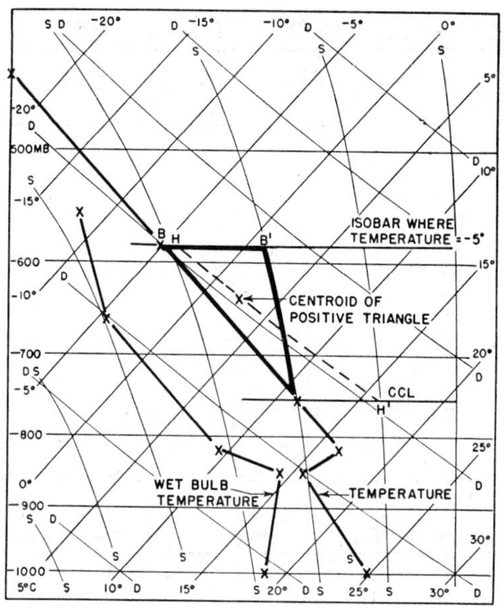

FIG. 10-1 Examples of evaluation of sounding for hail size. BB' is the base of the positive triangle and HH' measures the altitude, in this case 6 and 21 respectively. Entering the Hail Graph, Fig. 10-2 with these values, one obtains one inch for the diameter of the hailstones. [Skew T-log p Diagram, with mixing ratio curves omitted.] (from Fawbush and Miller, 1953)

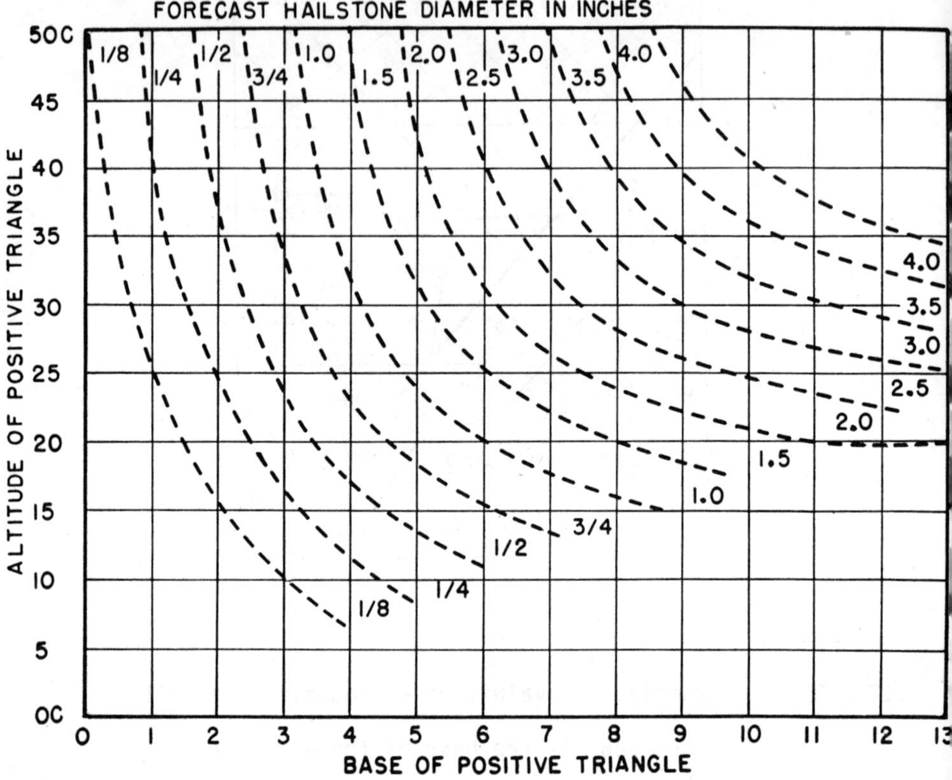

FIG. 10- 2 The Fawbush-Miller Hail Graph.

(from Fawbush and Miller, 1953)

and the isobar where the temperature becomes -5°C completes the triangle and serves as its base. This base is measured directly and its length in C degrees is the abscissa of the hail graph. The altitude is measured by tracing the dry adiabat from the base of the triangle, through the centroid to the CCL. The difference in temperature between the end points of this line segment is equivalent to the altitude in hectometers and is the ordinate of the hail graph. Entering the graph, Fig. 10-2, with these coordinates, 6 and 21 in the illustration, the size one inch is noted as the largest to be expected with this degree of instability."

The verification of 219 forecasts indicate that 77% of the cases show the validity of this method. The remaining 23% of the cases indicate reported hail size was either larger or smaller than forecast. However, as climatological data and newspaper accounts were used for this verification its accuracy is questionable.

Moreover, the development of these methods was largely empirical, and some of the parameters may be more related to the probability of a thunderstorm occurrence than to hailstone size. In addition, it is known that <u>not</u> all thunderstorms, even severe ones, produce hailstones.

10.2 Height of wet-bulb freezing above the terrain:

Fawbush and Miller (1953) have presented evidence to indicate a close correlation of the incidence of hail at the ground with the height of the wet-bulb freezing level above the terrain. A histogram of the number of reported cases of hail of various sizes by selected levels of 0°C wet-bulb temperature is shown in Fig. 10-3. The maximum hailstone diameter was reported to be greater than 7.5 cm when the wet-bulb freezing height was in the 7000 to 9000 foot (2.2 and 2.9 km) range. In nearly 96 percent of the representative soundings the wet-bulb freezing levels were between 5000 and 12,000 feet (1.5 and 3.8 km) above the surface. The probability of hail was found to be low when the wet-bulb lies outside this limit. Thus the height of the wet-bulb freezing seems related to hail melting.

Morgan (1970) examined the height of wet-bulb zero as a hail forecasting parameter in the Po Valley in Italy. His observations tend to raise doubts about the true efficacy of the wet-bulb zero as a hail predictor. He found a correlation between wet-bulb zero and low-level moisture. This suggests that low-level moisture may be the significant predictor, which can be directly measured from the sounding. Therefore, the wet-bulb zero might be a superfluous parameter.

10.3 Moisture in the lower atmosphere:

Appleman (1959), and Pappas (1962) emphasized

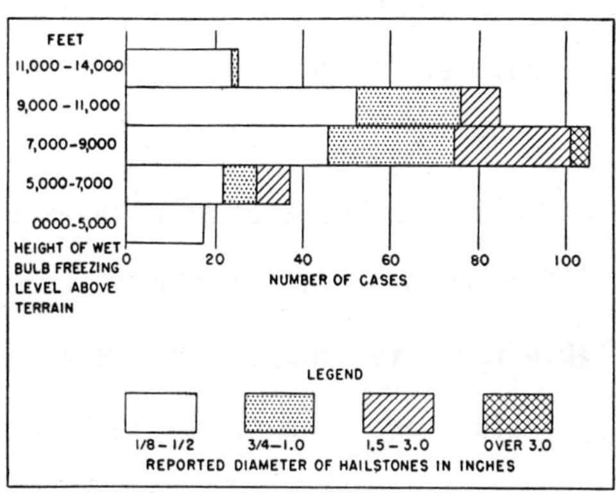

FIG. 10-3 Number of reported cases of hail of various sizes by selected levels of 0°C wet-bulb temperatures. (from Fawbush and Miller, 1953)

that when there is sufficient moisture in a cloud, hailstone growth is enhanced. Appleman (1959) reported a correlation between hail occurrence and the heights of the lifting condensation level and the melting level. If the cloud base is just 1 to 2 km below the melting level, hail occurrence is favored. However, the chances of formation are reduced if the cloud is too thick below the melting level.

The relationship has also been investigated by Longley and Thompson (1965) for Alberta hailstorms, though sufficient data were not available for this study. From radiosonde soundings the mean mixing ratios were computed at the 850 mb level for days of major hail, for days of minor hail and for days without hail. The mean dew points were 4.5°C, 5.3°C and 3.8°C respectively. Thus there was no clear distinction on the basis of moisture at 850 mb level.

The surface moisture was calculated regularly from the reported dew points. However, no satisfactory relationship between this parameter and occurrence of hail could be found after scrutinizing the data in several ways.

10.4 Cloud base temperature:

Wood (1955) reported the relationship between hail occurrence and cloud base temperature. The latter parameter was deduced from the convective condensation level and the positive area from the skew T-log p diagram indicating instability. The cloud base temperature seems to be strongly correlated with the occurrence of hail (Fig. 10-4). When the cloud base temperature is between 5°C and 15°C, hail occurs readily with relatively slight instability. When it exceeds 20°C, hail occurs rarely and with extreme instability. The demarcation line in Fig. 10-4 shows the occurrence of hail and non-hail thunderstorms.

10.5 The freezing level:

Pappas (1962) suggested a technique of hail forecasting, employing as predictors (1) the ratio of cloud depth below the freezing level to the cloud's estimated vertical development and (2) the height of the freezing level. Seventy severe convective storms in Texas, Oklahoma, Kansas and Nebraska (34 hail producing and 36 non-hail producing) were examined with respect to the above predictors. The scatter diagram is shown in Fig. 10-5. The predictor failed Pappas in 12 out of 70 instances. Although some association with these predictors has been shown, it is rather vague; Pappas does not explain the physical basis for this association.

10.6 Strong winds at high altitudes:

For the discussion on this predictor, the reader should refer to the sections on 'wind shear' (Sections 7.2.6 and 8.10).

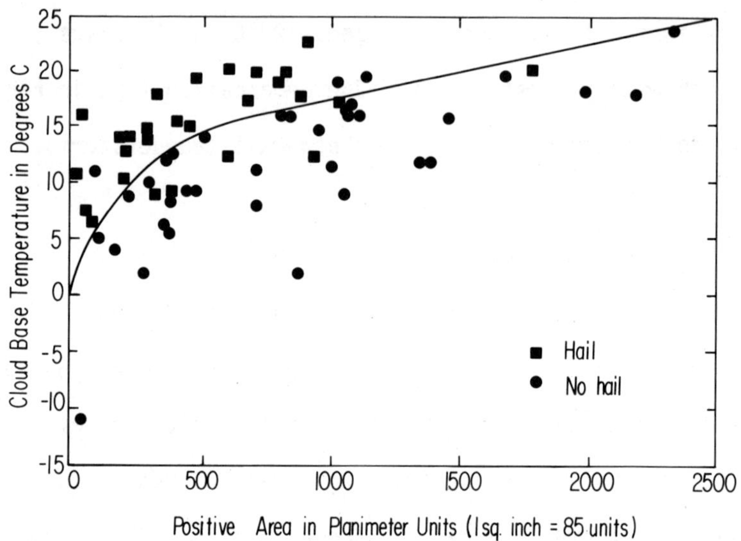

FIG. 10-4 Occurrence of hail and non-hail thunderstorms as a function of cloud-base temperature and instability. (from Wood, 1955)

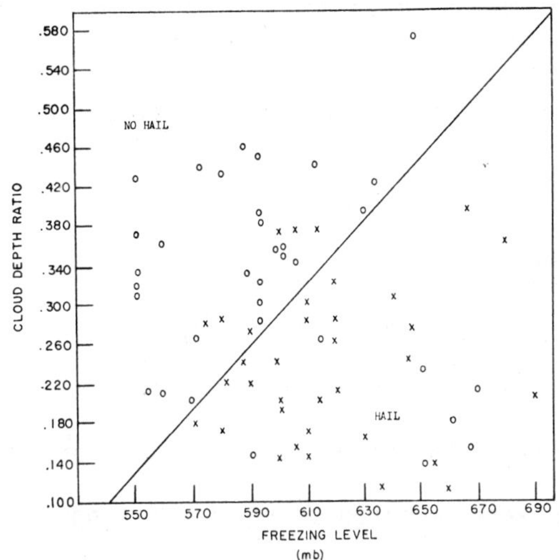

FIG. 10-5 Scatter diagram showing the distribution of selected hail occurrences at nine Midwestern stations during the spring and summer of 1959 and 1960. The freezing level height is plotted against the cloud depth ratio. [X = Hail reported. 0 = No hail reported.] (from Pappas, 1962)

10.7 Maximum height of radar reflectivity and the intensity of radar reflectivity:

For the discussion on these two parameters, refer to Sections 6.1 to 6.3.

10.8 Recent studies:

Longley and Thompson (1965) made an analysis of the incidence of hail in Southern Alberta during the years 1959-1963. The meteorological variables discussed earlier and studied by other scientists were examined to try to distinguish days with hail from those without hail. The presence of one or more parameters discussed in Sections 10.1 to 10.7 was found to increase the probability of hail. Yet it has also been shown that no one of these parameters is apparently necessary for hail to occur. Hence Longley and Thompson were forced to conclude that the margin between hail and no hail is apparently very small; their study has been unable to isolate this margin.

10.9 Public hail forecasting service:

The public hail forecasting service in the United States is provided by the National Weather Service's National Severe Storm Forecasting Center located at Kansas City, Missouri. The Air Force's Severe Weather Warning Facility provides a service for the military similar to the one that the National Weather Service provides for civil interests. Together they represent the largest concentration of professional experience in existence today in the forecasting of severe local storms and hail in the United States. The facility issues "watches" for severe thunderstorm activity, including tornadoes, large hail and damaging winds, over a defined area for a limited time period.

At present, no attempt is made to pinpoint the exact time and location of large hail occurrences. As in the case of tornado forecasting, a prediction is made of large hail occurrences somewhere over a defined area, say 160 by 320 km, during a specified period. Such a prediction does not carry any implication regarding the aerial extent of the hail, large or small.

The technique as described in Forecasting Guide No. 1 (Winston, 1956) is quoted below:

> "A technique for predicting hailstone size has recently been developed at the SELS Center (Foster and Bates, 1956) and is in current use. The mechanics of this method are quite similar to the empirical procedure of the Air Force Severe Weather Warning Center (Fawbush and Miller, 1953). The SELS method was derived under the following assumptions: a) hail size is proportional to the updraft velocity within a thunderstorm; b) the updraft velocity results from parcel buoyancy; and hence, c) the updraft may be approximated from positive area

measurements on a thermodynamic program. The use of this technique is based upon forecasts of severe thunderstorms made in conjunction with considerations other than positive areas or thermodynamic indices alone.

"Entrainment has been neglected here. In the case of the intense thunderstorms which produce large hail sizes, the parcels may be affected but little by entrainment at the core of strong updrafts of large cross-section where the larger stones are visualized to form. However, the spectrum of hail sizes observed over an area may be due to a spectrum of entrainment rates in a complex of thunderstorm cells. Undoubtedly, entrainment must be responsible for a considerable number of failures experienced with any hail forecasting technique which does not take its effect into account (and none in application do). In general, it should be expected that the effect of entrainment in reducing the size of the positive area would be greatest in the case of widely scattered thunderstorm activity, and least in an area of numerous, intense thunderstorms, such as along pronounced instability lines. There is no basic observational material on which to base a definite evaluation of the effect, however, and no operationally feasible procedure for including it at present.

"The hail size is estimated from a measure of the positive area on a thermodynamic diagram (WB form 1147). This positive area usually approximates a triangle and the measurements used here are H (height of the triangle) and ΔT (base of the triangle in degrees C). Available data point to a temperature of -10°C as the mean level of hail formation. Thus, the height of the triangular positive area, H, is measured from the level of free convection (LFC) to the -10°C level on the predicted parcel lapse rate. The base of the triangle, ΔT, is the temperature difference between the parcel and its environment. That is, the algebraic difference, -10°C minus the temperature of the actual environment or forecast sounding (forecast environment), at this same level. For this value to be significant, the parcel must be warmer than the environment. In those cases where the positive area is not approximately a triangle due to irregular observed or predicted temperature lapse rate, a smooth line is drawn to represent the mean lapse rate (equi-areal bisection). Actually, in working with prognostic soundings, such cases would not usually appear.

"An overlay, shown in Fig. 10-6, is used to

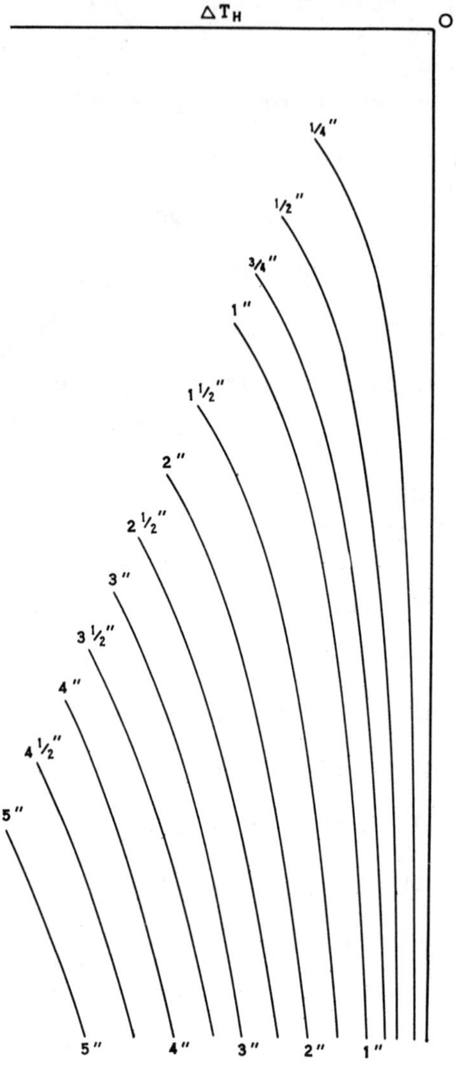

FIG. 10-6

Overlay for pseudoadiabatic chart for estimating hail size. Zero point is placed at -10°C point on parcel temperature curve with the abscissa (ΔT_H) along the pressure level of the -10°C intersection point and the ordinate (H) along the -10°C isotherm. Hail size estimates made from this figure must be corrected in Fig. 10-7 if the level of the -10°C parcel temperature is different from 400 mb. This figure is scaled so that it may be traced and used directly as an overlay on WB forms 1147 A, B, or C. (from Winston, 1956)

evaluate the positive area. This overlay is placed with the zero point (upper right) coincident with the -10°C point on the parcel temperature curve and the coordinates parallel to the appropriate coordinates on the WB 1147 A, B, or C base. An initial hail size estimate is read at the intersection of the isotherm corresponding to the environment temperature at the -10°C parcel temperature level and the isobar corresponding to the level of free convection. If the level of the -10°C parcel temperature is different from 400 mb (and it usually is) a correction for air density effect is made from the chart shown in Fig. 10-7 , which is based upon calculations using a -10°C parcel temperature at 400, 500, and 600 mb. The intersection point of the hail size initially obtained from the overlay (entered along the sloping solid lines) and the actual pressure of the -10°C parcel temperature level (entered along the ordinate) is extended vertically to the top of this graph to obtain the corrected hail size. It will be noted that the higher the actual pressure of that level, the greater the hail size.

"An example (graphical) of the use of this technique on a hail proximity sounding is shown in Fig. 10- 8."

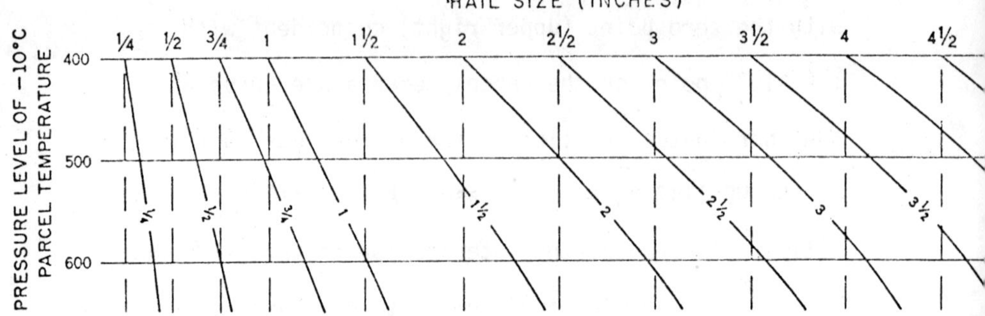

FIG. 10-7 Correction graph for cases where parcel temperature of -10°C occurs at a level other than 400 mb. This graph is entered along the sloping line corresponding to the hail size obtained from Fig. 10-6. The corrected size is read vertically above the intersecti of this sloping line and the pressure of the -10°C parcel temperature. Note that the higher the pressure the greater the hail size. (from Winston, 1956)

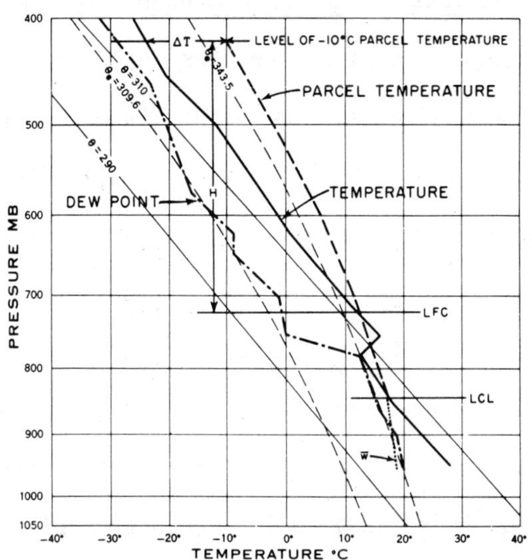

FIG. 10- 8 Upper air sounding at Altus, Okla., April 29, 1954, 1500 CST. Hail size was computed to be 2 1/4 inches. Hail up to 2 inches diameter fell 95 miles north of Altus, Okla. between 1530 CST and 1830 CST. (from Winston, 1956)

10.10 New suggestions to identify thunderstorms with hail:

To summarize, then, one has to conclude that the parameters discussed earlier may indicate the occurrence of a severe thunderstorm but cannot successfully distinguish thunderstorms from hailstorms. Instability is a major factor in causing hail, but the presently published data cannot identify it clearly with the occurrence of hail. For example, Appleman (1959) showed that great instability alone is not a sufficient condition for hailstone formation, although it does appe[ar] to be very closely related to the occurrence of thunderstorms. Wood (1955) and Appleman (1959) related the occurrence of hail and nonhail thunderstorms as a function of cloud base temperature and instability. Fawbush and Miller (1953) with some success correlated the occurrence of hail with the height of wet-bulb freezing and with large positive areas on the tephigram between the environmental dry bulb curve and the parcel adiabat. But other investigators like Douglas and Hitschfeld (1959), Longley and Thompson (1965) could not get the same correlation for their Alberta data. Though the maximum height of radar reflectivity was reported to correlate well with hail occurrence (Douglas and Hitschfeld, 1959), Geotis (1963) concludes from his radar data in New England hailstorms that there is no clear dependence of either hail size or duration of the radar reflectivity upon the height of the storm. The intensity of radar reflectivity, is a good indicator of heavy precipitation, but cannot distinguish between heavy rain and hail. The author's model of hail growth discussed in Chapter 8, indicates that not all of the above mentioned parameters are important. On the other hand, two parameters which were not considered by earlier investigators seem to be important for the formation of hail. These parameters are (i) height (Z_1 to Z_2) of the -4°C to -30°C temperature region, and (ii) duration for which the intensity of the radar reflectivity exceeds a minimum critical value in the region Z_1 to Z_2.

10.10.1 Occurrence of hail:

Atlas (1963, 1964) has reviewed the earlier work on the radar reflectivity factor and hail. Comparing the results of different investigators in different regions (Donaldson [1961] in New England, Inman and Arnold [1961] in Texas, and Wilk [1961] in Illinois), Atlas (1963) concludes that reflectivity alone is not a valuable indicator of the severity of a storm. But if we associate the intensity of radar reflectivity with the region and duration of its occurrence, we can identify hail or no-hail storms. The Gokhale and Rao (1969a) model indicates that the region of hail growth mainly lies between the temperature range -4°C to -30°C, the region between the lines Z_1 and Z_2 in Fig. 10-9. If a radar reflectivity factor higher than the critical value persists for several minutes in this region, the probability of occurrence of hail is rather high. Donaldson (1961) showed with his 3-cm data at 30,000 feet that 50 percent of the storms which produce hail larger than 1/2 inch have radar reflectivity factors greater or equal to 45 dBZ. Geotis (1963) reports that if the 10-cm radar reflectivity factor of a storm equals or exceeds 55 dBZ and these values are maintained for several minutes, then in New England at least, it can be identified as a hailstorm. Thus the values of 45 dBZ for 3-cm radar (not correc for attenuation) and 55 dBZ for 10-cm radar may be considered as minimum critical values for identifying hailstorms. If the high reflectivity value persists only below the height Z_1 or above the height Z_2, it does not indicate the formation of hail.

10.10.2 Nomogram to estimate hailstone sizes:

Prediction of the maximum size of hailstone that can occur in a hailstorm is as important as predicting the occurrence of hail itself. A nomogram has been developed (Gokhale and Rao, 1972) relating the size of hailstone with cloud base temperature, radar reflectivity, effective liquid water content (ELWC) and

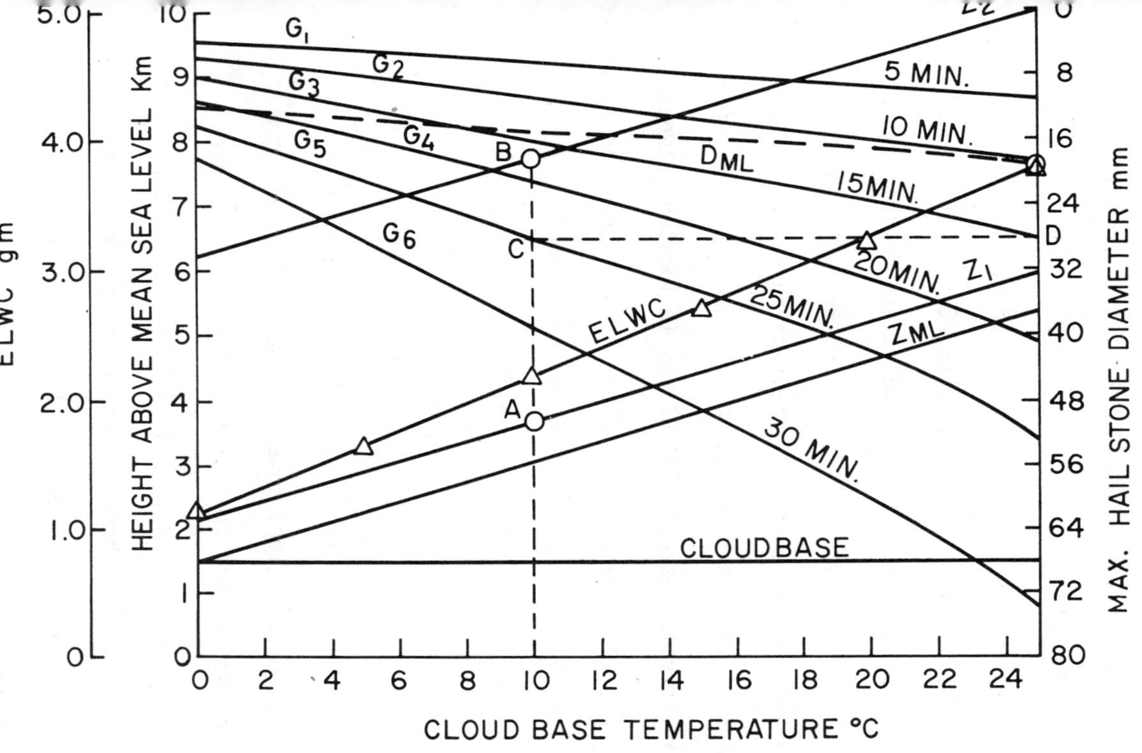

FIG. 10-9 Nomogram for determining the size of hailstone for cloud base height of 1.5 km. ELWC is the effective liquid water content corresponding to the indicated cloud base temperatures. Z_{ML} is the height of the melting level. (from Gokhale and Rao, 1972)

duration. The nomogram in Fig. 10-9 corresponds to a cloud base height of 1.5 km. For other cloud base heights, appropriate nomograms can be developed.

The nomogram consists of hail growth curves for different values of effective liquid water content (ELWC) and duration. It should be noted that the growth of the hailstone during its fall and its reduction in size due to melting and evaporation in the subcloud layer are not considered in these calculations. The following example illustrates the use of the nomogram. If 10-cm radar reflectivity value exceeds 55 dBZ and is located between the levels A and B for a duration of 25 minutes, proceed as indicated by the dashed line in Fig. 10-9. The maximum diameter of the hailstones for a cloud base temperature of +10°C, is 28 mm.

10.10.3 Conclusions:

The survey of available literature on radar data and the occurrence of hail lend some support to the author's (1970, 1972) suggestion that the maximum value of the radar reflectivity factor should lie in the -4°C and -30°C temperature region and should exceed a threshold value for a minimum duration. The probability of the occurrence of hail will be less if the maximum value of the reflectivity factor lies only above the height -30°C or below -4°C; it is also less if its value is less than 45 dBZ for 3-cm radar and 55 dBZ for 10-cm radar. This particular technique of forecasting hail limits the forecast time to 10 to 15 minutes and cannot be of much importance for forecasting hail well in advance, but still it may be adequate for hail suppression work.

The radar data presently published lack simultaneously observed information on all the three parameters, namely the maximum value of the radar reflectivity factor, the height of its occurrence (the temperature range at which the maximum reflectivity factor occurs) and the duration of its occurrence. In addition, the 3-cm radar data which is widely available is subject to ambiguity especially because of the problem of attenuation; one has to interpret the results with caution. The vertical temperature structure, derived usually from the radiosonde ascent nearest to the storm location and time, may not be a proper representation of the atmospheric conditions prevailing at the time of hail growth. These two facts and the lack of sufficient time variation studies of radar reflectivity factor call for a new investigation of thunderstorms with nonattenuating 10-cm radar and a well located radiosonde station. The radiosonde ascent should be programmed to cover the usual time of occurrence of thunderstorms at that location. Such simultaneous observations of the above three parameters and of the maximum size of the hailstones collected at the ground level are essential to draw firm conclusions on the occurrence and short range forecasting of hail.

Recently accurate radar observation of the hailstorm and precise observation of hailfall at the ground were correlated by Goyer (1971). The magnitude and location of the maximum radar reflectivity (at 10 cm) in the region where the temperature was between 0 and -30°C were observed. The time of hailfall, which was accurate only up to ± 5 minutes, was reported by ground observers or by telephone survey. The results of these studies are shown in Figs. 10-10a and 10-10b (the dotted lines correspond to the periods when hail was observed at the ground) and are summarized below:

i) The value of 10 log Z_e greater than 50 was noted for 30 minutes or longer for all hail producing storms.

ii) Hail was observed at the ground when the values of 10 log Z_e were between 50 and 66 and were between -4 and -31°C levels.

iii) No hail was observed at the ground for values of 10 log Z_e between 47 and 56 in the temperature range -20°C to -40°C.

Thus, these preliminary results confirm the author's suggestion that a relationship may exist between the hail size and the time the maximum radar reflectivity greater than 10 log Z_e equal to 55, spends in the temperature range between -4°C and -30°C. These findings are sufficiently interesting to warrant further study.

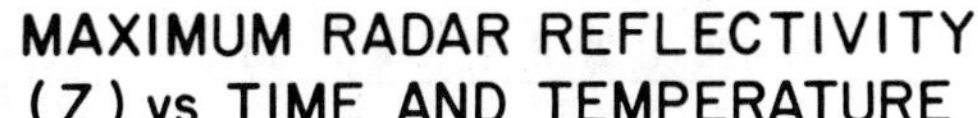

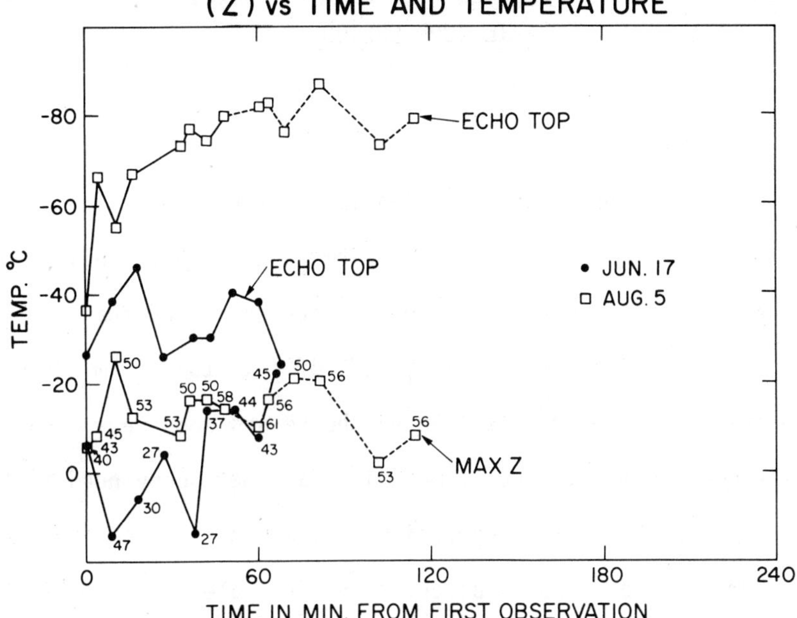

FIG. 10-10a (from Goyer, 1971)

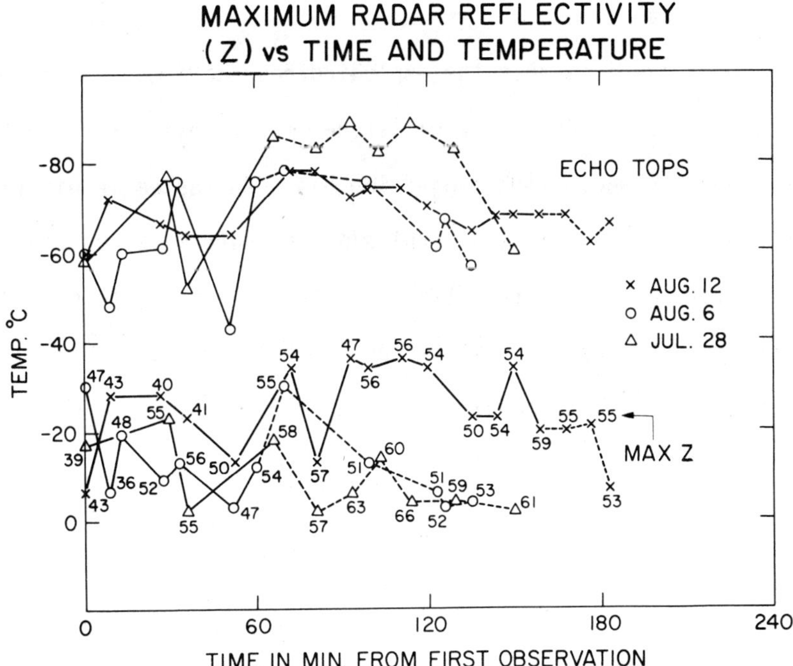

FIG. 10-10b (from Goyer, 1971)

CHAPTER 11

HAIL SUPPRESSION

11.1 Rocket firing:

Attempts at hail mitigation date back more than a century. In northern Italy some farmers have shot small rockets at hailstorms. The rockets carried a 1-kg charge of TNT which was detonated at altitudes of 1 km to 1.5 km. Simple and inexpensive materials like cardboard and wood were used for their construction. The farmers in Italy were convinced that the rockets are useful in suppressing hail and have fired thousands of them each year. Some observations indicate that the hailstones are softened by the rockets and, as a result, shatter on impact with the ground. This effect is claimed to be very localized in space and time.

The idea that the exploding nose cone may cause a shock wave capable of making the hailstone 'mushy' was first put forward by Vittori (1960). The suggestion has been tested in his laboratory by examining the effect of explosion pressure waves on the structure of hailstones. He attributed the observed effects to cavitation phenomena accompanying the passage of pressure waves of high energy through the liquid contained in the hailstones.

However, List (1963b) repeated Vittori's experiments and reported negative results. Hollow spheres filled with ethylene dichloride and snowballs of low mechanical cohesion drenched in water did not show any effect when exposed to blast waves. Artificial hailstones consisting of ice-water mixtures and grown in a hail tunnel also indicated negative results. Therefore List concluded that there is no justification for attempting to combat hail by exploding rockets.

Favreau and Goyer (1967) studied the effect of shock waves on ice cubes in the laboratory. The cubes were weakened when exposed to shock especially if the cubes contained a water column. The weakening of the ice cubes was ascribed to shock-induced cavitations within the water columns. Of course, ice cubes very crudely simulate hailstones. Moreover, shock-tube experiments failed to produce these effects in hailstones. Bubbles and cracks have been observed within hailstone samples as a result of the re-freezing of the liquid phase even in the absence of shock waves.

Subsequent examination of the hail suppression work in Italy using explosive rockets failed to show any significant reduction in hailfall (Morgan, 1973). The opinion of the farmers seems to have changed and relatively few are being fired nowadays. The communes which were purchasing those rockets during the epoch of their high fashion, slowly withdrew from these activities (Gori et al., 1971).

During the early sixties little was heard about rocket firing, except Sansom (1965) reported similar work in Kenya, Africa. The anti-hail rockets used in this project contained 800 gm of TNT in an explosive head. These were designed to burst at 1.5 to 2.0 km above ground level. During a period of half an hour, 50 to 150 rockets were fired in a single storm. The firing started only after hail started falling and it continued until hailfall subsided. According to Sansom the firing of explosive rockets did reduce damage caused by hail, but he pointed out the necessity of more work along these lines before definite conclusions could be drawn.

To date no one has studied any of the so-called 'mushy' hailstones following rocket firings, and hence this effect has not been conclusively established.

11.2 Hail suppression concepts:

Different ways have been suggested of attacking the problem of hail suppression. The storm could either be overseeded to freeze most of the supercooled water, or seeded with condensation nuclei to increase the efficiency of the collision and coalescence process. Another suggestion is to increase the number of embryos and thus reduce their final sizes. Let us consider each of these suggestions in detail.

11.2.1 Overseeding the storm:

The growth of large hail is due to accretion of a large

amount of supercooled water. One way of diminishing growth would be to reduce the supercooled water reaching the hailstones and thereby reduce the accretion rate. Since ice nuclei can convert supercooled water to ice, the supercooled water droplets would be changed to ice crystals if sufficient number of nuclei are introduced into the storm. This process, known as overseeding the storm, may affect at least a major portion of it. To achieve overseeding, Ludlam (1958) suggested the formation of 100 ice nuclei per cm^3, while Weickmann (1953, 1964) suggested one ice nucleus per cm^3.

The principle seems to be theoretically sound, but it will be necessary to see whether it is economically feasible to introduce the large quantities of AgI material that will be required. A cubic meter of cloud may contain 10^8 droplets. If the core of the hailstorm is 3 kilometers in diameter and an average updraft in such a developed cloud is at least 10 m/sec, then 6×10^{17} droplets would be moving up the core each minute. A silver iodide burner may produce 10^{13} nuclei effective at -10°C per gm of AgI. If it is further assumed that all particles are reaching the cloud and are uniformly distributed, about 6×10^4 grams or about 130 pounds of AgI would be used in one minute. If the storm lasted for at least a couple of hours, a large amount of silver iodide would be required and the cost would be prohibitive at $50 per pound of AgI. Therefore, it does not seem practical to try to mitigate hail by overseeding.

11.2.2 Seeding with condensation nuclei:

Weickmann (1964) suggested that adding great amounts of condensation nuclei at the base could retard the growth of large droplets due to coalescence. If hailstones grew only from large liquid drops that have formed in the lower part of the cloud, as suggested by Ludlam (1958) and Appleman (1959), the above suggestion might have merit. But it is now well recognized that hailstones grow not only from supercooled large drops but from graupel as well. Seeding as described above will not have any effect on the reduction of the number of graupel particles which originate out of tiny supercooled drops and ice crystals.

11.2.3 To increase the number of ice embryos:

The **third method** is to increase the number of ice embryos so that the size of the hailstones will be correspondingl reduced. **This is based on the fact that**

$$\frac{4\pi}{3} R_a{}^3 \; N_a = \frac{4\pi}{3} R_b{}^3 \; N_b \qquad (11\text{-}1)$$

or,

$$R_a = R_b \left({}^{N_b} \big/ {}_{N_a} \right)^{1/3} \qquad (11\text{-}2)$$

where the subscripts a and b indicate numbers after-seeding and before-seeding,respectively, N is the number of ice embryos and R the radius of the hailstone. Theoretical work indicates that the water content of a hail cloud becomes effectively depleted by a small number of hailstones (about 10 per m^3), so that increases of their concentration by two orders of magnitude can be expected to decrease their size sufficiently to prevent damage. Thus, producing many more embryos by AgI seeding, perhaps by droppable pyrotechnic flare system, should yield smaller hailstones, which would be less damaging and more likely to melt before reaching the ground (Summers et al., 1972).

It should be pointed out, however, that the competitive concept in hailstone growth would work provided the concentration of embryos is sufficiently high. Theoretical calculations are needed to determine the critical number. For smaller concentrations, more hailstones of the same size would result and not more stones of smaller size, as direct competition would predict.

Another possibility is that the seeding of the hail cloud with AgI at levels where the in-cloud temperatures are between -5°C and -10°C will freeze a large number of millimeter size water drops by contact nucleation. Glaciation will liberate a large amount of latent heat of fusion which will then increase buoyancy and lead to the intensification of the already strong updraft. This will result in taking the hail embryos to higher levels where the temperature is lower than -30°C. In this situation, the water droplets carried upwards from the base will naturally freeze before reaching the hail embryos and reduce their further growth.

The following studies indicate the effectiveness of the contact nucleation mechanism as compared to other nucleation mechanisms in freezing large millimeter size drops.

11.3 Droplet freezing by contact nucleation:

Droplet freezing by contact nucleation has been observed (Gokhale and Goold, 1968) during the course of studies of ice nucleation in the author's laboratory. Apparatus to provide a constant rate of cooling and a specially constructed cold chamber were used for this study. Silver iodide particles sprinkled on supercooled millimeter-size water drops were found to be effective in freezing the drops at -5°C. Particles of naturally occurring silicates were effective in the range -7 to -10°C. Other conclusions of these studies are summarized below:

1) Particles freeze water drops much more effectively when they make contact at their surfaces than when they are embedded in the drops. The shift toward warmer temperature is in the range of 5-10°C depending on the material.

2) Particles affect the nucleation of water drops much more when the drops are supercooled before the particles land on their surface than when they are not. The effective degree of supercooling is -5°C for silver iodide and -8°C for silicate material.

3) Among the materials tested, AgI particles are the most effective by contact nucleation. This shows that a surface structure of the particle identical with that of ice from the point of view of crystal symmetry and lattice dimensions are important in this mechanism.

4) Particles of micron-size seem to be more effective than submicron particles in freezing drops by contact nucleation.

5) As the number of particles contacting the surface of the drop increases, the probability of its freezing at a warmer temperature increases.

Silver iodide particles embedded in millimeter-size drops are not effective in freezing the drops at a median temperat of -15°C, as shown by Hoffer (1961) and Gokhale (1965b).

Microcinematographic studies of contact nucleation were also carried out in the author's laboratory. Water drops of 2 mm diameter were supercooled on a glass slide, sprayed with AgI particles, and photographed at low magnification with a 16-mm movie camera at a speed of 64 frames per second to determine the nature of freezing by contact nucleation (Gokhale and Lewinter, 1971). In each case, nucleation was initiated at the point of particle contact and continued over the entire surface of the drop. This shell of ice then froze toward the center at a much slower rate. The surface structures and crystallization rates by contact nucleation compare favorably with those found by other workers where water was frozen in containers. Other conclusions of this study are listed below.

1) An AgI particle contacting the surface of a drop does not penetrate it but remains on the surface. The freezing is initiated at the point of contact.

2) The time for freezing to begin after a micron-size particle touches a drop which was supercooled to -5°C ranged from 16 to 47 msec.

3) Nucleation may commence from more than one particle simultaneously.

The third study involved the freezing of freely suspended, supercooled drops by contact nucleation (Gokhale and Spengler, 1972). Water drops were balanced in the updraft of a large vertical wind tunnel and allowed to supercool to the ambient temperature. Ice crystals introduced into the updraft were, as expected, the most effective nucleants in freezing the drops at ambient temperatures of 0°C and below. The results of this experiment using silver iodide and clay as contact nucleants closely agree with the findings from earlier work performed with a constant rate of cooling apparatus. The effective temperature for 100 percent nucleation efficiency in the case of AgI particles was -4° to -5°C; for silicate particles it was in the range -7°C to -10°C.

Thus, the experimental results on the freezing of supercooled water drops by contact nucleation demonstrate that drops freeze at much warmer temperatures by the contact nucleation mechanism than by other ice nucleation mechanisms. Hence, best results will be obtained in freezing millimeter-size water drops in clouds by contact nucleation if silver iodide particles are introduced in the region where the degree of supercooling is at least -5°C. In a hail-producing cloud, such a seeding should result in a large number of frozen drops that could prevent further growth of hailstones.

11.4 Hail suppression projects:

During the last decade a considerable effort has been made in several countries towards the suppression of hail. It was believed that the knowledge of the trigger mechanism of cloud seeding would be useful in reducing the hail damage. Most projects of hail suppression (except for the Russian project) use an indirect method of silver iodide seeding from cloud base and ground generators, aircraft flares, or else use small rockets containing AgI that are dispatched against the storm in general. Important projects were undertaken during the last decade in the countries listed below:

UNITED STATES (Schleusener, 1962, 1968)
FRANCE (Dessens et al., 1950 and 1967)
SWITZERLAND (Sänger, 1960; Schmid, 1967)
BAVARIA (Müller, 1964)
ARGENTINA (Grandoso and Iribarne, 1961)
KENYA (Henderson, 1968)
RUSSIA (Sulakvelidze, 1968)

Attempts were made to increase the number of hail embryos by seeding clouds with silver iodide or lead iodide. Except for the Russian effort, all the large scale field programs were set up even though no physical model of a hailstorm had been developed. Russian scientists have formulated certain ideas as to how hail forms in clouds, the details of which have been discussed in Chapter 7.

11.4.1 Projects in the United States:

In the United States over the last two decades very few systematic experiments were designed and carried out to check hail suppression feasibility. But a large number of experiments of cloud seeding aimed at reducing the damage to crops by hail have been carried out by commercial operators. Very little reliable information is available in the scientific literature regarding these projects. It seems that the same techniques used in increasing rainfall by seeding with silver iodide generators are used in most of these hail suppression projects. It is known that the intensity and occurrence of hailstorms are much more variable and unpredictable than are those of rainstorms. Therefore, the effect cloud seeding has on hail suppression is much more difficult to determine. Many of these experiments did not have adequate controls to discriminate between natural and artificial effects.

The survey and history of hail suppression operations in the United States during 1949 to 1957 have been given by Frank (1957). Most of the operators used ground based AgI generators for seeding. A few projects utilized aircraft for seeding operations, and tracked clouds by radar. Routine weather watches were maintained throughout all projects. A sufficient number of silver iodide nuclei was introduced into a potential hail-producing system in order to limit the growth of hydrometeors so that they would precipitate out in the form of many showers of small particles.

The evaluation techniques for the success of hail suppression projects employed by commercial operators were essentially crude. The tests used were not statistically rigorous. From the client-contractor relationship, the most impressive result is shown in Fig. 11-1 where a steady growth in target area is shown from 1949 to 1957, with a projection to a still greater area in 1958. This does not indicate definite success per se; however, projects were deemed sufficiently worthwhile to be continued in subsequent years. The same conclusion was drawn from the hail insurance statistics. Though these tests were not satisfactory, the farmers were eager to do something to offset their recurring losses and were inclined to accept the physical evaluations based on less crop damage.

Schleusener (1968) summarizes the results of hailstorm seeding projects and suggests certain generalizations regarding seeding rates. The locations of various domestic projects were in Colorado, Wyoming, Nebraska, South Dakota and North Dakota. Using modern techniques, a high concentration of seeding material was injected with greater accuracy of placement with airborne equipment. One of the main conclusions of the findings was that "evaluation of experience to date supports the hypothesis

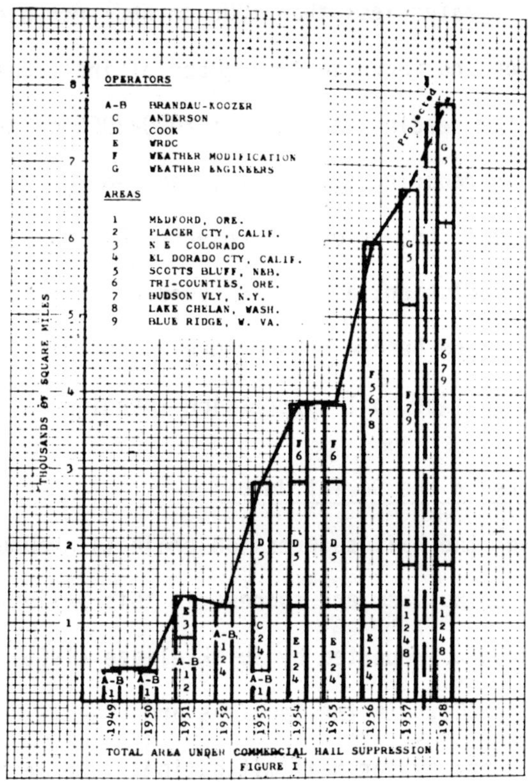

FIG. 11-1 A steady growth in target area is shown from 1949 to 1957 and a projection to still greater area in 1958. (from Frank, 1957)

that seeding at rates less than 1000 gm hr^{-1} per storm may stimulate convection and increase the number of individual hail events, but that heavier seeding at rates of 2000-3000 gm hr^{-1} per storm is effective in reducing the total impact energy from hail falls."

However, the evidence presented by Schleusener does not support this conclusion, and it is necessary to obtain more experimental evidence before drawing overly optimistic conclusions about the efficacy of hail suppression techniques.

11.4.2 Projects in the Soviet Union:

Among the hail suppression projects around the world, the greatest enthusiasm and success in this activity comes from the Soviet Union as reported by Sulakvelidze (1966a) and by Battan (1969, 1970). In the Caucasus and Transcaucasus, large programs are underway for hail mitigation. The method is to identify precisely the region in a convective cloud where the hail is beginning to form and then to inject AgI particles in that precise region by means of artillery shells and rockets. An adequate statistical evaluation is lacking in their experiments. However, empirical evidence such as decrease in crop damage has convinced the Russian experimenters that their technique is highly successful. All these projects have reported spectacular success in reducing hail damage. In the protected areas the damage was 3 to 5 times smaller than that in the unprotected area. The cost amounted to 2 to 3 percent of the value of the crops saved.

11.4.2.1 The High Altitude Geophysical Institute:

The work in hail suppression at the High Altitude Geophysical Institute (Known as VGI) was directed by G. K. Sulakvelidze and has been reported in a number of publications. The two main ones are the books by Sulakvelidze, Bibilashvili and Lapcheva (1965) and by Sulakvelidze (1966a). Besides this activity, there are other programs underway in the Soviet Union in hail suppression - in Georgia, Armenia, Moldavia, and Central Asia. But these have not been so well reported.

There are in all 16 radar units and 82 artillery pieces in the VGI operation. About 500 people are involved in this activity, out of which 20 to 25 are scientists. The period of hail suppression activities extends from May to October. The ice nuclei are introduced into the particular regions of the clouds by means of artillery (Gusev, 1966) and the accuracy in firing is satisfactory, the explosions occurring within 300 m to 400 m of the desired spot.

Sulakvelidze (1966b) described the conditions which were thought to be necessary in order to conduct a successful hail suppression program and these are summarized by Battan (1970):

> "1) the ice nuclei must be dispersed in the 'hail center' in a volume not exceeding 15 km^3;

"2) the time interval between the radar detection of the hail center and the introduction of the nuclei must not exceed 2-3 min;

"3) the emptied reagent containers must not contribute a hazard to the populace;

"4) the time intervals between introductions of the ice nuclei must not exceed 2 min;

"5) the amount of reagent (lead or silver iodide) dispersed at any one time should be about 100 g; and

"6) to modify all clouds within a given area there should be 3-4 points from which the reagent can be delivered."

During the period from 1962 to 1965 some 650 experiments on hail suppression were carried out in the Northern Caucasus and Transcaucasia. The effect of treatment was apparent after 2-3 minutes, indicated by radar observations. An attempt was made to protect some 250,000 hectares in 1965; a comparison with reference sectors indicated that the hail damage was only one-ninth the usual damage in this protected region.

The results of the Caucasus Expedition's operations over two years for the protection of cultivated areas against hail are presented in Table 11-1. They indicate that hail damage is reduced by 50 percent according to long term averages. The hail damage in the protected region was less than one-fourth

TABLE 11-1

Summary of work performed by the Caucasus Expedition on the protection of cultivated areas from hail damage. (from Sulakvelidze, 1966a)

	Protected sector			Reference sector		
Years	Area under crop, thousand hectares	Damaged area under crop, thousand hectares	Ratio of damaged to total area, %	Area under crop, thousand hectares	Damaged area under crop, thousand hectares	Ratio of damaged to total area, %
1959--1963 (no hail control measures taken)	260.0	20.3	8	247	20.0	8
1964	292	15.8	5	247.0	22.1	9
1965	303	9.7	3.1	312.0	59.1	19
Mean for years of protection	298	12.7	4.2	279.5	40.6	14.5

of that in the reference region. The performance in protection was improved considerably in 1965; this improvement is attributed to more experience gained during earlier years and fewer mistakes made during this period.

The experience gained in 1964 and 1965 showed that "the best basic anti-hail unit is a detachment consisting of one radar point and four guns, which can protect a cultivated area of 100,000 to 120,000 hectares from hail damage. Taking a two-year average, the mean number of projectiles required for the protection of 100,000 hectares under crop is 670 per season."

In addition it demonstrated that

"1) The method of hail control developed by VGI can be recommended for adoption on a national scale for protection of large areas from hail damage, since its adoption reduces the damaged areas by an average of 75 %.

"2) The aerological-synoptic method of hail forecasting has an average accuracy of over 90 %, which not only confirms the correctness of the concept of the mechanism of hail formation on which it is based, but also recommends it for adoption on a national scale.

"3) The method of locating hail centers by radar, using multiwavelength equipment, makes it possible not only to determine the boundaries of the hail nucleation zone, but also to estimate the diameter of the hail reaching the ground, with an accuracy of 35%.

"4) The nonsplintering "El'brus-II" projectile, when used in thickly populated regions with intensive hail processes, fully satisfies the demands, i.e., conveying the reagent into the hail center without any danger of splinters to the population.

"5) The efficient organization of hail prevention work was evolved.

"6) The cost of protecting 1 hectare of cultivated land amounts, on the average, to about 1 ruble, 50 kopecks."

[1 hectare = 2.47 acres; 1 ruble and 50 kopecks $\simeq$ 2 U.S. dollars]

11.4.2.2 Institute of Geophysics of Georgian Academy:

Kartsivadze (1967, 1968) of the Georgian Academy of Science describes his operation consisting of 40 rocket stations for firing lead iodide into potential hailstorms. The number of damaging hailstorms which occur on about 60 hail days a year was reportedly reduced by 70%. They occur most often in May and June.

Battan (1969) gives the following account from his personal visits to these projects:

"At the main field station there is a

rocket launcher which can hold four missiles. There is a plotting and control room to which radar information is transmitted by telephone and plotted on a map. The controller decides the time and place of rocket launching.

"A crucial instrument for this entire operation is a 3 cm radar set. Operators continually scan the "protected" area and make a number of measurements of the echoes either approaching or over this area. On the basis of the measurements of each individual cloud, decisions are made regarding the probability of hail in the cloud. If the maximum radar reflectivity measured as log $Z < 2$, it is concluded no hail is present. When log $Z \geq 2$, eight other radar parameters are used to decide the probability of hail. They are the following: 1) the depth of the radar echo; 2) the height of the region of maximum radar reflectivity; 3) the temperature at the top of the echo; 4) the temperature at the height of maximum radar reflectivity; 5) the height of the top of the echo; 6) the depth of the region of high echo intensity. (This is defined as that region where the echo intensity is 10 decibels below the maximum value of echo reflectivity.); 7) the temperature of the level of the top of the region of high echo intensity; and 8) the ratio of cloud depth above and below the 0C isotherm. On the basis of the study of past data, regression analyses have been made of the probability of hail as a function of these eight parameters. A small analog computer for obtaining the probability of hail has been built. In any particular circumstance, the eight input parameters are punched on a keyboard and the probability of hail in the cloud appears on a digital counter. . . . .

"In the Alazani Valley, 460,000 hectares are "protected" by the rocket network. The area is divided into three polygons. There is a total of 12 rocket sites spaced at intervals of 15 to 20 km. Some rockets are 10 to 12 km upwind of the "protected" area. Each rocket has a maximum range of about 9.5 km. In practice, an attempt is made to fire the rocket into the cloud at about the -5C level.

"The rocket being used this year is called the 'Oblakov' and is in a launcher which can handle four rockets. A new rocket launcher capable of holding twelve rockets is being brought into service. The lead iodide used as a seeding reagent produces 10^{12} nuclei per gram at -10C according to laboratory tests. A new rocket called the 'Alazani' is now under development and will have different ballistic properties from the 'Oblakov.'

"The 'Oblakov' is about 2 m long and about 15 cm in diameter. The propulsion system is in the center of the rocket, and the power is applied for a period of a few seconds, and then the missile continues on a ballistic trajectory. The forward third of the rocket contains the seeding reagent. It usually is lead iodide but might be silver iodide. The quantity of the seeding reagent, in the form of a pyrotechnic mixture is 3 kg per rocket, and it burns out in 45 sec. The plastic nose of the rocket unscrews, and a very simple dial adjustment is made to indicate the range from the launching pad at which the pyrotechnic seeding mixture becomes ignited. At the back third of the rocket is a chamber for a parachute. Another simple control is used for setting the range at which the parachute will open.

"In practice, the radar information is telephoned to the control room, the radar data is used to calculate the probability of hail, and, if the probabiltiy of hail exceeds 5 %, the controller prepares to fire. He obtains radar data and plots it on the board showing the rocket stations and the "protected" area. From the plotting board, the controller reads azimuth and range from the potential hailstorm to the nearest rocket station. Radiosonde data indicates the altitude of the -5C level. The controller reads off the following coordinates to the appropriate rocket station: azimuth angle; elevation angle; the range at which the pyrotechnic mixture is to be ignited; and the range at which the parachute is to be opened. This information is transmitted via radio-telephone. The people at the rocket station, consisting of a crew of at least two, fire as soon as they can get the rocket and the launcher adjusted. At the central station in the Alazani Valley, the time required from the reading of the coordinates to the firing of the rocket was certainly less than a minute. The actual firing of the rocket is done inside a small building some 30 meters from the launcher. Apparently the electrical firing signal is obtained by depressing a plunger. . . .

"Various analyses have been made of the efficacy of the technique of using rockets to seed hail clouds. Over a seven-year period within the "protected" area, hail frequency was measured at 32 stations. It was found that, during the seeded period, hail frequency was 70 % less than during the non-seeded period.

"A second series of evaluations was made by comparing crop damage over the target area and over a nearby control area. The statistical data on crop damage over a 32-year period was compiled. This was done for both the target and control areas. This information was used to predict an expected amount of damage over the target on the basis of observed damage over the control area. The expected damage then was compared to the observed crop damage during the seeded period. The damage statistics were obtained from agricultural agencies concerned with crop insurance. According to Dr. Kartsivadze, since seeding started in 1961, the annual decrease of crop damage caused by hail ranged from 61 to 80 % and averaged 79 %. He reported that they calculated a benefit-to-cost ratio of about ten to one, making the operations well worthwhile.

"On this project in the Alazani Valley, there are about 53 research people of various kinds, and some 250 others who operate radar equipment, the rocket launchers and do other routine jobs. The latter group are paid for, and supplied, by the agricultural interests in the 'protected' area."

11.4.2.3 Moldavian:

Procedures similar to those above were employed in Moldavia by scientists from the Central Aerological Observatory (Voronov et al., 1967 and Battan, 1969). They used rockets which carry 5 kg of pyrotechnical mixture and 3.1 kg of lead iodide. To track the storms and to make echo measurements, 3.2 cm radar was used from which the probability of hail was estimated. It was reported that during 1966 hail damage was reduced to a tenth of that of previous years.

Battan (1969) gives the following account from his personal visits to these projects:

> "The Moldavian Anti-Hail Expedition carries out the seeding operations over an area of 215,000 hectares by means of 19 rocket stations some 10 to 12 km apart. They employ the same type of rocket, the "Oblakov," which is being used by Dr. Kartsivadze's group in the Alazani Valley. Lead iodide is the seeding agent. The maximum range of the rockets is about 9.5 km. . . . .
>
> "Incidentally, Dr. Gaivoronskii also directs an anti-hail operation over 80,000 hectares in the Crimea.
>
> "In Moldavia where the most prominent crop is grapes, about 90 % of the hailstorms are associated with frontal systems. The point frequency of hail is 4 to 6 days per year. In the region, as a whole, there are perhaps 31 hail days per year.
>
> "There are between 75 and 80 people involved in the entire operation. Of these, 57 work at the 19 rocket launching sites.

Before 1966, the program at Korneshti was mostly concerned with research on the properties of hailstorms and on hail-seeding techniques, but since 1966 the hail seeding has been on an operational basis.

"Six radar parameters are used to determine the likelihood of hail. It might be recalled that in the Alazani Valley eight parameters are employed. . . . . A small analogue computer similar to one used by Dr. Kartsivadze's organization was constructed allowing the operators to enter the 6 radar parameters and have an output reading of hail probability.

"Dr. Gaivoronskii indicated that the cost of operations in Moldavia was between 400,000 and 500,000 rubles a year and that the benefits over the approximately 200,000 hectares amounted to about 2 million rubles, leading to a benefit to cost ratio of 4 or 5 to 1."

11.4.2.4 Transcaucasian Hydrometeorological Research Institute:

Personnel from the Transcaucasian Hydrometeorological Institute in Tbilisi, Georgia use the method (Bartishvili et al.,

1967) of firing ice nuclei into regions where temperatures are between -7 and -15°C. Hygroscopic particles are also dispersed at the levels where the temperatures are between 0 and -7°C. The introduction of hygroscopic particles is to encourage the formation of precipitation and the initiation of downdraft.

Battan (1969) describes the techniques used in these experiments:

> "The procedure used to seed hailstorms by scientists of the Transcaucasian Research Institute includes the use of artillery to insert ice nuclei into the supercooled part of the cloud. It also includes the dispersion of large salt nuclei into the warm part of the cloud. This is being done by means of artillery shells containing salt.
>
> "Some 500-700 grams of common salt is introduced into the cloud for every cubic kilometer of cloud. This is intended to produce some 400 to 600 particles per cubic meter, with the particles being five to ten microns in diameter. The object of this aspect of the seeding is to wash out the precipitation from the lower part of the cloud. It follows a scheme similar to one suggested some years ago by Herbert Appleman and by Frank H. Ludlam.

By seeding with ice nuclei in the upper part of the cloud, it is hoped to produce more and smaller hail particles which would melt before they reach the ground. This technique of using both ice nuclei and salt is not based primarily on producing more hailstones. . . . . The liquid water content in cumulonimbus clouds could be higher than 5 to 6 gm m^{-3} and that an accumulation zone might exist as droplets are carried up from below the region of the updraft maximum. There seemed to be some reluctance in being too specific on what was regarded as a realistic maximum liquid water content in hail-producing thunderstorms. The point was made that in an air mass thunderstorm, axes of updrafts might be nearly vertical, but, in frontal conditions with substantial environmental wind shear, the axes of the thunderstorms might be tilted.

"The Transcaucasian Hydrometeorological Research Institute is responsible for seeding hailstorms over an area of 150,000 hectares southwest of Tbilisi.

"During the operational program, hailstorm forecasts are made 12 to 24 hours in advance. Radar is used to obtain data on the presence of hail and hailstone sizes. . . . . It was stated that the seeding began when a cloud was developing, the maximum radar reflectivity (log Z) exceeded a certain level and the temperature at that level was between -1 and -8C. It was said that if the seeding started early enough, hail was prevented from forming. . . . . The technique for evaluating the effectiveness of hail suppression involves comparison of hail damage over the target area and that over a nearby area. In addition the character of the radar echoes over the two areas are compared. The two areas are about the same size and have the same physical geography."

11.4.2.5 Hail suppression in 1972:

From 18 July to 31 August 1972, J. D. Marwitz from the University of Wyoming, U.S.A., was an official scientific exchange visitor to the U.S.S.R. and reported the results of the latest in a series of scientific exchanges concerning the Soviet anti-hail program. His main findings are summarized below (Marwitz 1973):

(a) The North Caucasus region resembles northeastern Colorado from a topographical and meteorological point of view. However, the lines of storms are aligned <u>with</u> the wind rather than the normal <u>to</u> the wind.

(b) Large supercell and multicell storms do occur in the Caucasus region, but are rare. The Soviet scientists are apparently unable to suppress all the hail from these types of storms.

(c) The personnel in the Georgia Academy of Sciences indicated that 100 to 200 ice crystals per liter are needed to seed a cloud, and since only 0.1% of the particles injected actually act as ice nuclei, about 10^5 particles per liter must be injected.

(d) The single-stage rockets have a range of 5 to 6 km, while the two-stage rockets have a range of 4 to 10 km, depending on the elevation and fuse setting.

(e) The hail damage was decreased by 60 to 80% and the rainfall was increased by 10 to 20% during seeding years.

11.4.3 Argentine project:

The aim of the five year Argentine project was to determine whether the AgI seeding of convective clouds would result in significant hail suppression and diminish the damage to vineyards in the Mendoza area (Grandoso and Iribarne, 1963; Iribarne and Grandoso, 1965). The experimental area chosen for the project was 4000 km^2 of flat country just to the east of the Andes, between 68° and 69° W and between 32°40' and 33°20' S. A dense network of observers from the vineyards, covering an area of over 1200 km^2, was established. The results were evaluated on the basis of the reports from this network.

The seeding was done with the help of generators located on the ground, which used charcoal impregnated with an acetone solution of AgI. These generators produced about 6×10^{13} nuclei per gram of AgI at -15°C. The burners were all operated by the farmers.

About 100 generators were in operation, out of which 20 were on the eastern slope of the Andes and the remaining 80 in the other experimental area as shown in Fig. 11-2. The generators were laid out in rows in a north-south direction. The rows were 10 to 15 km apart and the generators were placed 4 to 5 km apart in every row. The prevailing winds in the lower layer were perpendicular to the direction of the rows.

In Argentina, the hail season extends from October to March. About 30 hailstorms on an average occur per season. A method of randomization of the seeding days over the area was utilized. During the project period a total of 96 control (non-seeded) days and 102 seeded days resulted.

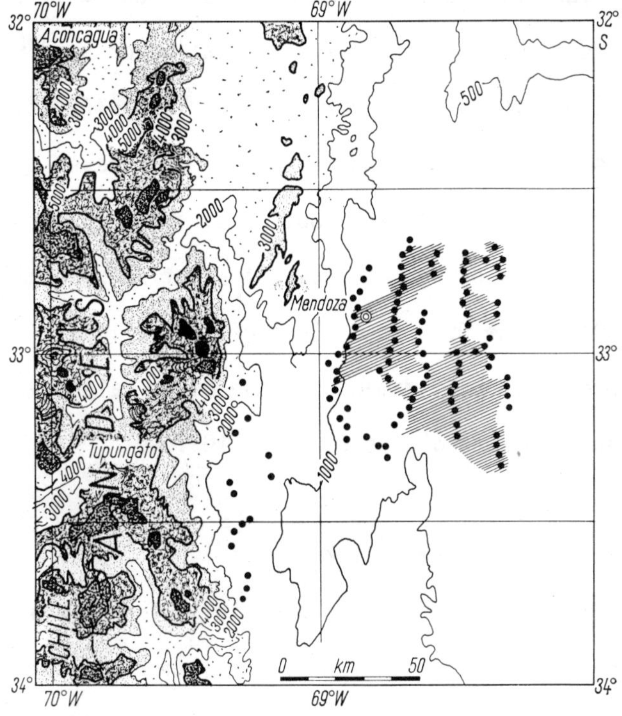

FIG. 11-2 EXPERIMENTAL AREA. The shaded area represents the cultivated part, mostly covered by vineyards, in the experimental area. Black dots represent seeding posts. Figures on lines of constant altitude are in meters. (from Grandoso and Iribarne, 1963)

The hail-damage parameter was essentially the product of hail area and intensity. The evaluation of seeding led to inconclusive results. However, when the classification of weather situations as a cold or a warm front or 'air mass' type was made, more interesting results were obtained. When cold front situations were seeded, hail damage decrease of about 70 percent was significant at the 0.10 to 0.15 significance level. An increase of 100 percent in hail damage for the other category was significant for seeded days at the 0.10 level.

11.4.4 Project in South-Western France:

The French effort started in 1951 on a small scale under the scientific direction of the late Professor H. Dessens. The object was to suppress large, damaging hailstones in the Aquitain Basin of southwestern France. The project has been gradually increasing until now. The effort is well organized and well recognized in the rural districts of France. In 1970, 350 ground-based generators were used to treat an area of 7 million hectares. The program extends over a period from April to October of each year. Each burner uses 9 gm hr^{-1} of AgI in acetone and runs about 10 hours each day and 100 times in a year.

J. Dessens (1968) reported a reduction of hail damage for the seeded period as compared to earlier periods, but the result was inconclusive because the total period of operation was not long enough. After four more experimental seasons, the results have been reported by J. Dessens and J. Lacaux (1972) using the same economical factor but a new statistical basis. The target area and the control areas in this project are shown in Fig. 11-3.

During the eight years 1963 to 1970 the insurance crop-loss data in the target area was compared with the loss in the control areas; it showed a consistent and significant decrease of the hail damage in the target area.

An opposing view on the French program is offered by Boutin et al. (1970). Their statistical studies on French operations indicate that the seedings have either no effect or an effect too low to be statistically significant.

11.4.5 Swiss project:

A large hail suppression project was carried out in Switzerland from 1959 to 1962 (Sänger, 1960; Sänger et al., 1962). Seeding was strictly randomized and was done with AgI ground generators. A weather forecast for the following day was received at 1630h. If it were to be a day with possibilities of thunderstorms, a randomized decision determined whether seeding was to be done. The seeding started at 0730h and ended at 2130h. The ground-based generators were pulsed 5 min on and 10 min off, with a rate of consumption of AgI of about 800 grams per day. Hailfalls occurring only during the period between 0800h and 2200h were used for statistical evaluation. After

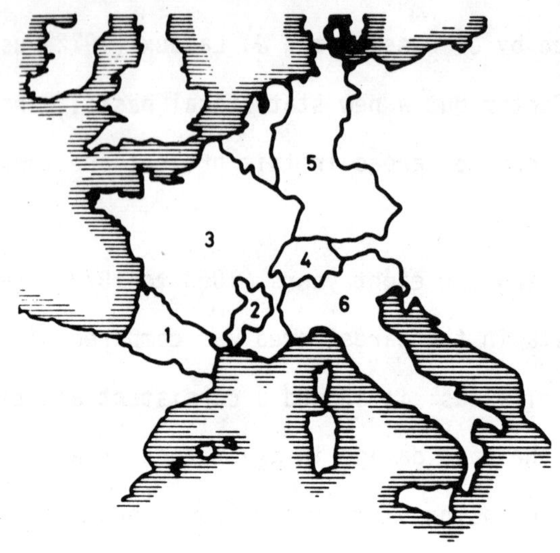

FIG. 11-3 The target area [1] and the control areas [2, 3, 4, 5, 6]. (from Dessens and Lacaux, 1972)

four years of operations, evaluations indicate very disappointing results for the hail suppression experiments. As a matter of fact, there was an increase of 68 percent in hail days in the seeded areas over the unseeded areas, at a significance level of 0.04 (Schmid, 1967).

11.4.6 Project in Bavaria:

On the northern slope of the Alps, the region called the Bavarian Plains is frequently affected by hail fall. An experiment using silver iodide released from rockets as well as from ground generators to test the possibilities of hail suppression in this area has been described by Müller (1967).

The intention of the experiment was to suppress hail as much as possible in the target area, which was chosen to be the Rosenheim district. Hailfall in Bavaria is a rather infrequent event and is also restricted in area. The most severe hailstorms are characterized by strips of hailswaths 10 to 20 miles wide and up to several hundred miles long. These were usually parallel to the chain of the Alps and to the prevailing wind direction. The investigation to determine the main features of the occurrence of hail damage consisted of checking the official weather reports and the reports of the official hail insurance organization.

The map (Fig. 11-4) shows the Rosenheim district surrounded by the other districts of Ebersberg, Aibling,

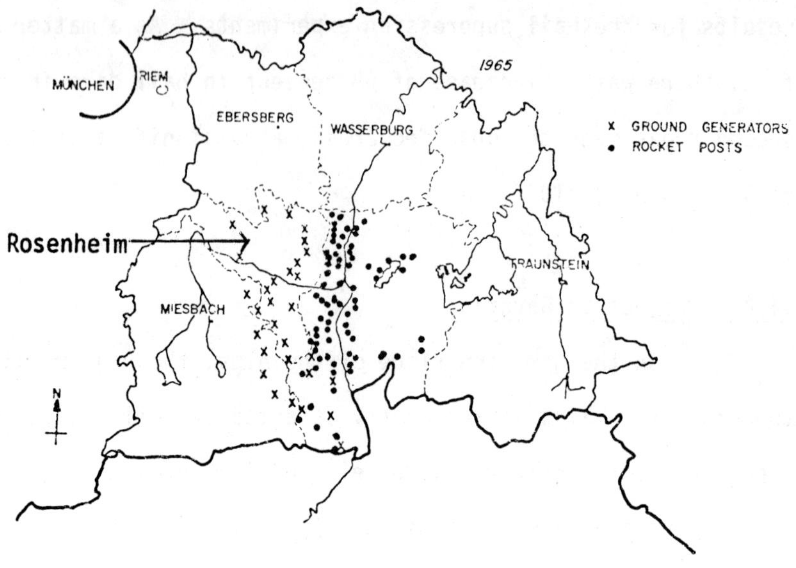

FIG. 11-4 Rocket posts and ground generators in Bavaria, 1965. (from Müller, 1967)

Miesbach, Wasserburg (control area A), and Traunstein. The locations of the rocket posts and the ground generators are also indicated in the map. The total number of rocket posts installed was seventy-six, each post having ten rockets at its disposal. Each rocket contained 800 gms of cheddite and 16 gms of AgI. The rocket had the capacity to attain a maximum height of about 1.5 km above the ground. However, the freezing level in summer over Bavaria is about 3 to 3.5 km and hence, all the seeding material might not have reached the region of maximum effectiveness. In addition to rockets, thirty propane gas generators dispersing 1 gm of AgI per minute in a solution of NaI and acetone were utilized.

Though randomization is the only basis for statistical evaluation, this experiment could not be carried out on a random basis because of practical difficulties. The results of the experiment were inconclusive. The number of townships damaged by hail multiplied by the number of days with damage was summed and recorded for each year and each district. The sum of the hail damage days in the target area (Rosenheim) during the experimental period from 1958 to 1965 was compared with that in the preseeding period from 1950 to 1957. The damage occurred on 72 percent of the seeded days and on 100 percent of comparable days during the preseeding period. The numbers for the control area A were 71 percent and 100 percent respectively. Thus, there was less natural hail damage during 1958 to 1965 than in the period from 1950 to 1957; therefore no effect of seeding appears to have resulted.

11.4.7 Hail suppression project in Kenya:

An operational hail suppression program near Kericho, Kenya was undertaken by Henderson et al. (1970). Approximately 45,000 acres of select tea are in production in the Kericho-Nandi Hills area of Kenya, East Africa. A general location map of the area is shown in Fig. 11-5. The tea estates lie at elevations between 1.8 and 2.0 km msl.

The design of this operational field program was decided from experience and data obtained during a number of field programs conducted in the United States during the 1956-'60 period. The design involves introducing AgI nuclei by aircraft seeding in the inflow areas. These are identified as important to the development of hailstones within each thunderstorm cell. The remaining larger portion of the storm which does not generate hailstones was left untreated. The rate at which material was released in the inflow areas produces at least 100 per liter ice nuclei concentrations effective at -15°C. Pyrotechnically generated AgI was used exclusively in this program.

The primary and secondary target areas of 800 sq. miles each are separated by some 40 miles. The locations of these areas and other geographical features are shown in Fig. 11-5. The areas of tea grown were too large and the storms too

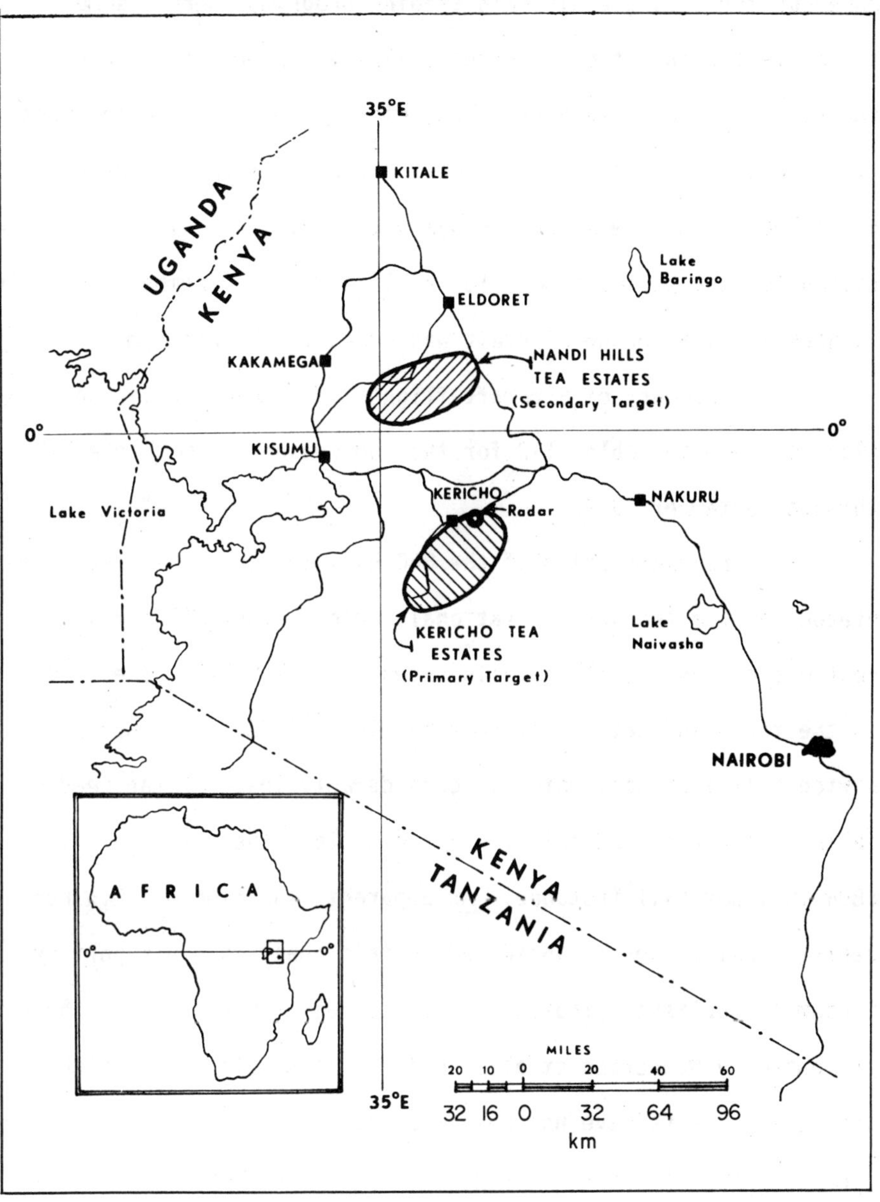

FIG. 11-5 Kericho-Nandi Hills hail suppression program. (from Henderson et al., 1970)

numerous for a single aircraft seeding program. Hence, some selective and operational criteria were necessary; for example, the Kericho area would have first priority for all seeding operations, and was treated as a primary target, whereas the Nandi Hills Tea Estates were the secondary target. This situation lasted for the period from October 1967 through December 1968. In January 1969, a second aircraft was added to the program.

A summary of aircraft flights and silver iodide application is given in Table 11-2 for the two seasons from October 1967 through September 1969.

The examination of the 860 hail instances has been completed for the two year operational period. The hail damage investigation reveals the details shown in Table 11-3. The values in the table indicate an average tea loss of 2863 lbs. per reported hail instance, which is considerably less for the seeded cells than non-seeded cells. In the latter, the tea damage is 6894 lbs. per hail instance. An apparent reduction of 58.5 percent in hail losses is indicated by this analysis. A report by Sansom (unpublished) indicates results less favorable than those presented by Henderson et al. Besides, it should be noted that statistical tests have not been applied to hail damage data. However, the results presented by Henderson et al. do show reduction in hail damage to the tea crop due to airborne seeding.

TABLE 11-2

Hail Suppression Program - Operational Summary

(from Henderson et al., 1970)

	1967/68	1968/69	Total
Operational days on which seeding flights were conducted	184	247	431
Total seeding flights	247	398	645
Seeding flight hours	268	381	649
Observation and other flights	135	200	335
Flight hours in this category	86	91	177
Pyrotechnic seeding devices (aircraft units)	1,764	2,264	4,028
Grams of silver iodide (aircraft units)	112,550	118,370	230,920
Pyrotechnic seeding devices (ground units - Nandi Hills)	3,379	2,265	5,644
Grams of silver iodide (ground units - Nandi Hills)	50,840	40,265	91,105

TABLE 11-3

COMPARISONS OF HAIL DAMAGE TO TEA

(seeded - vs - non-seeded)

(from Henderson et al., 1970)

	SEEDED CELLS		NON-SEEDED CELLS	
	Hail Instance	Tea Loss (lbs.)	Hail Instance	Tea Loss (l
1967-1968	58	169,900	358	2,552,500
1968-1969	165	469,000	279	1,839,200
TOTAL	223	638,900	637	4,391,700
AVERAGE LOSS	2863 lbs/hail instance		6894 lbs/hail instance	

11.5 Summary:

The principal point to be made in summary of the projects discussed in the earlier sections is that, except for hail suppression experiments in the Soviet Union, all other projects show rather inconclusive results. They indicate the possibilities that hail damage might be increased or decreased by hailstorm seeding.

The experiments in the United States using ground or airborne generators have been inconclusive. The results of the long term experiments carried out in Switzerland and France were similar. However, the latest reports claim some success in the French effort. Experiments in Argentina show both positive and negative results for different types of storms. The reports of the National Academy of Sciences published in 1966 therefore remark that "there is a wide range of opinion on whether or not hail can be effectively suppressed or its damage mitigated."

The Russian scientists are far more optimistic. They claim significant success because, with the help of anti-aircraft shells and rockets, they are introducing the silver iodide particles directly into that portion of the cloud where the water is supercooled and high in concentration. For various reasons, definite proof of nearly complete success of their experiments is difficult to accept at face value. Some independent verification is obviously needed.

To check the various techniques of hail suppression it is essential to design systematic and controlled experiments. Present technology has the potential to allow productive and meaningful research on hail mitigation. The scientific efforts, especially on hail modeling, need technological verification. Such an effort, namely the National Hail Research Experiment (NHRE) coordinated by the National Center for Atmospheric Research (NCAR), is now underway (see Appendix C).

APPENDIX A

THE COLLECTION EFFICIENCIES OF HAILSTONES

Langmuir and Blodgett (1946) have made theoretical calculations for the collection efficiencies of smooth spheres for small droplets. Macklin and Bailey (1966) have extended these calculations for smooth hailstones of various diameters as illustrated in Fig. A-1. The collection efficiencies for droplets of 15 μ radius fall from 0.9 to 0.65 as the stone increases in diameter from 1 to 6 cm. However, the value of the collection efficiency for droplets greater than 30 μ radius is almost independent of the hailstone size.

Macklin and Bailey (1968) have described experiments where a sphere or a hailstone was held fixed in an icing tunnel and its mass was measured after it had been exposed to the droplet stream for a short time. The collection efficiency (E) was determined directly from mass increase. However, the validity of the assumption of uniform droplet flux was uncertain. Hence, the ratio (E_1/E_0) was calculated from the ratio of mass increase for any two objects when the rate of spraying was kept constant. The median radius of the droplets used was 14.5 μ and the spraying rate was chosen so that the deposits grew near the wet limit. The ambient temperature was ≈ -10°C. In the first experiment, the tunnel speed was set at 30 m sec^{-1} and the

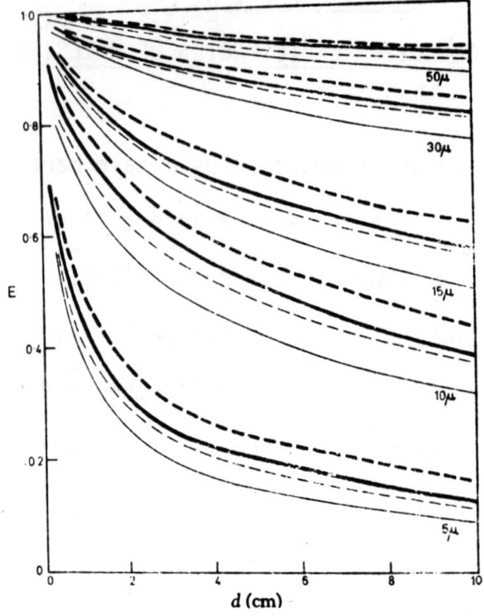

FIG. A-1 The collection efficiencies, E, of smooth spherical hailstones for cloud droplets of various radii are shown as a function of hailstone diameter, *d*. The full curves are appropriate to the -20°C level and the broken curves to the -40°C level. The calculations have been carried out using values for the drag coefficient of 0.55 (heavy lines) and 1.1 (light lines). (from Macklin and Bailey, 1966)

collection efficiencies were determined as a function of radius of the hailstone (Fig. A-2). In the second experiment, the variation of collection efficiency of 1.2 cm diameter ice sphere was determined as a function of airspeed (Fig. A-3). There was no significant difference between the values for the artificial hailstones and those for smooth spheres of the same cross sectional area. The results, however, were generally lower than those predicted theoretically by Langmuir and Blodgett (1946).

This may be due to the loss of accreted water by splashing and shedding during hailstone growth. Recently, Carras and Macklin (1973) reported the results of icing tunnel experiments on the shedding. The loss of accreted water occurs in both the dry and wet growths at ambient temperatures warmer than -12°C. The loss during dry growth is due to shedding of water due to the aerodynamic shear from the stress liquid layer which covers the surface of the growing hailstone and this would affect appreciably the growth rates.

Fig. A-4 shows the values of liquid water concentrating (W') assuming that no accreted water is lost, plotted against the actual experimental values (W). Values of W for which accretions were observed to enter the wet growth regime are indicated on the Figure. If the experimental collection efficiencies correspond to the theoretical values, the relation between W' and W in Fig. A-4 would be linear and the slope of the line would be unity. The

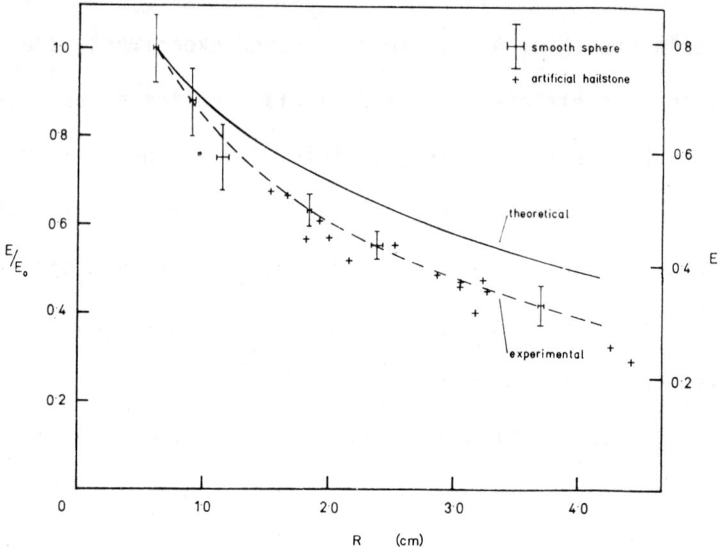

FIG. A-2 Variation of the collection efficiency of smooth spheres and artificial hailstones as a function of radius at an air speed of 30 m sec^{-1}. The scale on the left is the ratio E/E_0 calculated from Eq. (1); that on the right is the value of E calculated directly from the estimated liquid water concentration in the tunnel. (from Macklin and Bailey, 1968)

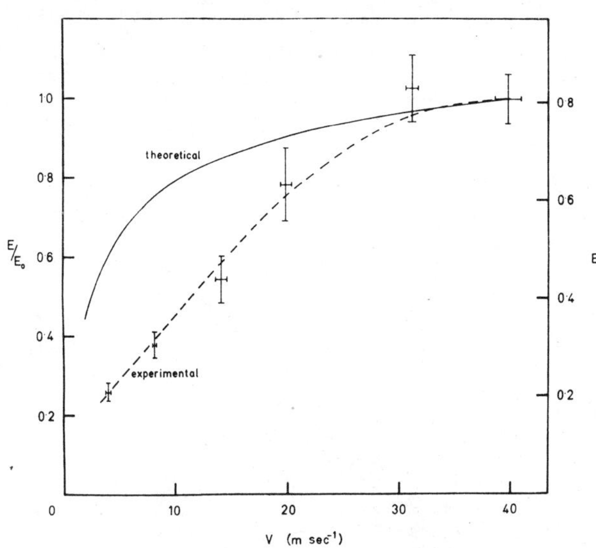

FIG. A-3 Variation of the collection efficiency of 1.2 cm diameter ice spheres as a function of airspeed. The scale on the left is the ratio E/E_0 calculated from Eq. (1); that on the right is the value of E calculated directly from the estimated liquid water concentration in the tunnel. (from Macklin and Bailey, 1968)

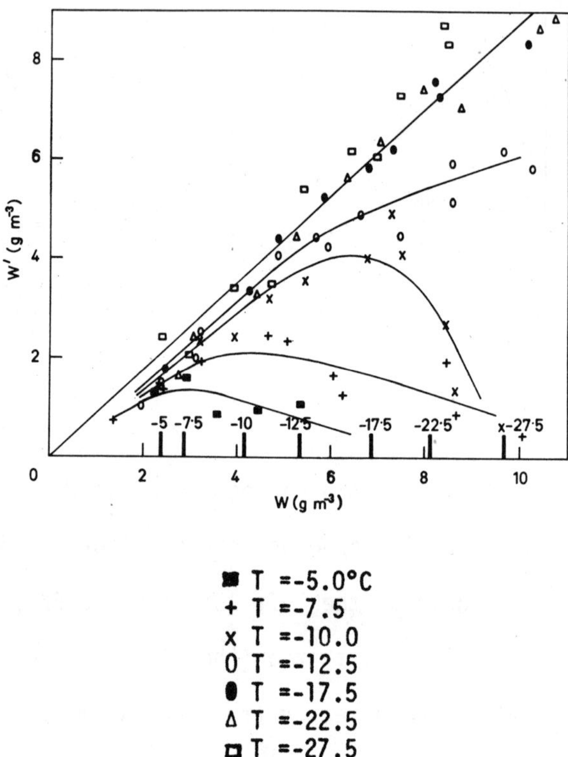

FIG. A-4 Values of W' calculated from Eq. (2) for the artificial hailstones plotted against W, the liquid water concentration in the tunnel. (Carras and Macklin, 1973)

relation is linear for temperatures at -17.5°C and below and the slope of the line is 0.90 for the artificial hailstones. At temperatures of -12°C and warmer the relation between W' and W is non-linear in both the dry and wet growths, indicating that some of the accreted water is lost.

APPENDIX B

THE MELTING OF HAILSTONES

The rate of melting of a hailstone falling through a still atmosphere depends on (a) the rate of transfer of heat by conduction and convection and (b) the rate at which latent heat is liberated by the condensation of water vapor upon its surface or loss of heat by evaporation.

Ludlam (1958) and Macklin (1963) in their treatment of the melting, make a realistic assumption that most of the melt-water is shed. Thus a hailstone is covered with a very thin film of water so that the surface temperature may be assumed to be 0°C. Following the treatment given by Ranz and Marshall (1953) and Macklin (1963) the rate of melting of a spheroidal hailstone, falling in still, clear air with its shortest axis vertical can be expressed by

$$\frac{dm}{dt} = \frac{\chi Re^{1/2} A\beta}{2r_H L_f} \qquad \text{(B-1)}$$

where m is the mass of ice spheroid or hailstone, χ is the (Fig. B-1) numerical factor in heat transfer coefficient whose value depends on the ratio of the axes of the spheroids, A is the surface area of the spheroid, Re is the Reynolds number, r_H is the semi-major axis of oblate spheroid or radius of sphere, and

$$\beta = Pr^{1/3} k\Delta T + Sc^{1/3} L_v D\Delta\rho_v \qquad \text{(B-2)}$$

where

Pr is the Prandtl number;

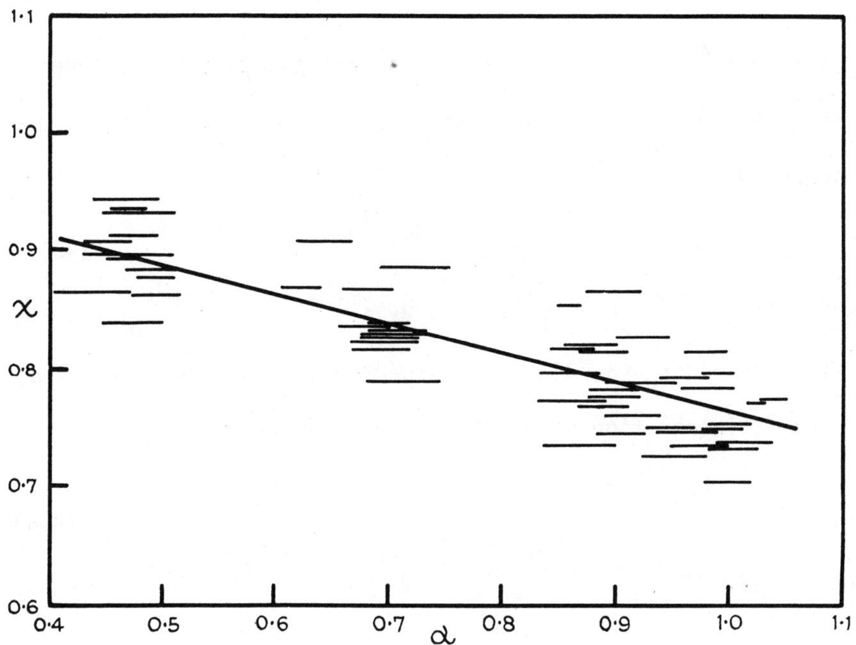

FIG. B-1 Dependence of the numerical factor χ in the heat transfer coefficient on the ratio of axes of the spheroids, χ. (from Macklin, 1963)

T	is the air temperature;
Sc	is the Schmidt number;
$\Delta\rho_v$	is the difference in water vapor density at surface of hailstone and airstream.

Since $m = \frac{4}{3}\pi\rho_i r_H^3\alpha$, where ρ_i is the density of hailstone; for constant α, which is a ratio of major to minor axes of a spheroid,

$$\frac{dr_H}{dt} = \frac{\chi Re^{1/2}A\beta}{8\pi r_H^3\alpha\rho_i L_f} \qquad \text{(B-3)}$$

Since $V^2 = 8gr_H\alpha\rho_i/3C_D\rho_a$ and $dr_H/dt = Vdr_H/dz$ and assuming $\rho_i = 0.$ g cm^{-3}, integration of equation (B-3) yields

$$r_{H_o}^{7/4} - r_{H_g}^{7/4} = 2.48\text{x}10^{-3}\chi\gamma C_D^{1/4}\alpha^{-5/4}\int_0^{z_0}\rho_a^{3/4}\beta\mu^{-1/2}dz \qquad \text{(B-4)}$$

where r_{H_o} and r_{H_g} are the values of r_H at the 0°C level and at the ground respectively, and $\gamma = A/4\pi r_H^2$.

The term $\chi\gamma C_D^{1/4}\alpha^{-5/4}$ in equation (B-4) indicates the effect of shape of the stone on its melting. The change in the maximum dimension during melting is more pronounced the smaller the value of α.

Atlas and Ludlam (1960) showed that the equation (B-4), for spherical stones falling through still air under conditions appropriate to the Wokingham storm, takes the form

$$r_{H_o}^{7/4} - r_{H_g}^{7/4} = 0.535 \qquad \text{(B-5)}$$

For various values of α, the value of r_{H_g} as a function of r_{H_o} is shown in Fig. B-2. The lowest curve shows that a spherical hailstone should have a radius larger than 5 mm at the 0°C level if it is to reach the ground in hot weather. If the radius at this level is appreciably greater than 5 mm, the reduction in size during the fall to the ground is not appreciable.

Bailey and Macklin (1968) have reported that the heat transfer from stones of radius > 3 cm and having lobed structure is three times that assumed in equation (B-5). Therefore for such stones equation (B-5) should read

$$r_{H_o}^{7/4} - r_{H_g}^{7/4} = 1.61 \qquad \text{(B-6)}$$

which suggests that even very large stones, if they are lobed, would undergo appreciable melting in the sub-cloud layer.

The equations B-1 to B-6 are for the melting rates of moderate and large sized hail; however, no experimental evidence has been offered to verify the correctness of these approximate equations.

Newman and Gokhale (1975) conducted experiments in which artificially produced hailstones of various sizes ranging from 1.5 to 3.5 cm in diameter were suspended by means of nylon thread in the updraft of a large vertical wind tunnel. The hailstones were allowed to melt for time intervals of up to five minutes. Melting

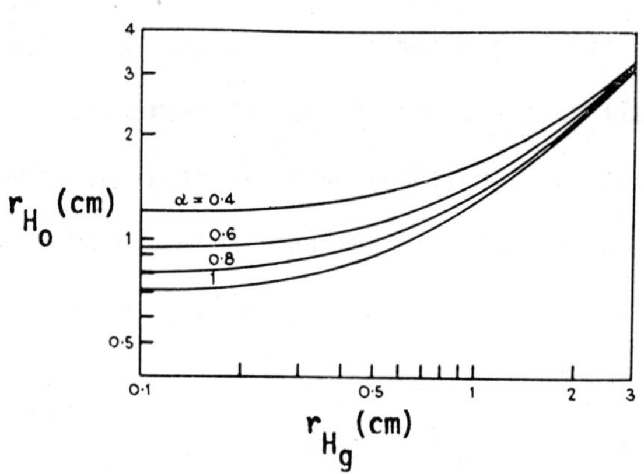

FIG. B-2 The value of the semi-major axis of a spheroidal hailstone at the ground (r_{H_g}) as a function of the value at the 0°C level (r_{H_0}) for various axis ratios. It has been assumed that the hailstones fell through clear still air. (from Macklin, 1963)

rates were calculated for one minute intervals and were found to be in excellent agreement with theoretical results. However, the value of χ, the heat transfer coefficient, was found to be 0.62 for spherical hailstones, as opposed to the value of 0.76 given by Macklin (Fig. B-1).

Taking into account the effect of melting, threshold diameters were calculated for three different atmospheric situations. Hailstones less than 0.6 cm in diameter will not survive a fall to 20C from 3000 m. Hailstones less than 0.78 cm in diameter will not survive a fall to 28C from 5000 m. This last value of 0.98 cm compares favorably with the threshold value of 1.16 cm found by Ludlam (1958) for hailstones falling from 5400 m.

It was also found that a hailstone of 1 cm diameter falling from 3000 m would reach the ground with a diameter of 0.75 cm after 210 sec. This is in sharp disagreement with Drake and Mason (1966) who calculated that a 1 cm diameter hailstone, falling in an updraft of 3.5 m/sec at a temperature of 10C would melt completely after 11 sec.

In addition, some interesting shapes were found to be related to the speed with which air moves past a falling hailstone. Stones falling at speeds greater than 8 m/sec relative to the air were found to become pronouncedly disk-shaped or flat-bottomed during melting. Stones falling at speeds less than 8 m/sec relative to the air were found to become conical-shaped during melting. These preliminary observations suggest that further study is needed to determine the effects of melting on hailstone shape.

APPENDIX C

THE NATIONAL HAIL RESEARCH EXPERIMENT (NHRE)

The purpose of the National Hail Research Experiment (NHRE) is to develop capabilities for the study and modification of hailstorms. It was designed to determine statistically if damaging hail was being suppressed by cloud seeding techniques and to assess the societal impacts and economic benefits. It is an intensive five-year project which started in the summer of 1972 and involves the expertise of private industry, government, and university research groups. The experiment is carried out in a 7,500 km^2 area of Northeast Colorado over and adjacent to the Pawnee Grasslands by the National Center for Atmospheric Research (NCAR) and primarily funded by National Science Foundation (NSF) in an amount exceeding 12 million dollars.

This is an area which produces an average of 20 hailstorms each annual three-month season. The following hailstorm features are to be continuously observed throughout the storm's life cycle: its synoptic and meso-scale environment; the three-dimensional profiles of its updraft velocities; liquid water content; radar reflectivity; and surface hail and rainfall. The basic observation systems consist of the following:

1. Aircraft (Sabreliner, Queen Air, DC-6, C-130, F-101B, Piper Aztecs and T-28 armor-plated

penetrating aircraft).

2. Sailplanes (the NCAR/NOAA Explorer and LS-1 from the University of Chicago, etc.)
3. Three radars (the NHRE dual-wavelength (10 cm and 3.2 cm) radar, the University of Chicago/ Illinois State Water Survey (CHILL) dual-wavelength radar, and the Desert Research Institute (DRI) M33 radar were used for the acquisition of real-time radar information in addition to their research functions).
4. Dropsondes.
5. A meso-scale meteorological network which included four radiosondes and 22 surface meteorological stations.
6. Eight mobile survey crews and 420 precipitation ground network stations.
7. Two mobile meteorological crews.
8. The NCAR Ice Physics Group to collect hailstone samples.
9. Cloud Photography.

The Federal Aviation Administration (FAA) aircraft control, Data Acquisition and Display Systems (DADS) and communications facilities are the basic support systems (NHRE summary reports, 1972, 1973, 1974).

The hail suppression operation begins with the declaration of a "hail day," which then becomes a seed or no-seed day on a random draw. AgI flares and vertically launched rockets are fired alternately if the prescribed threshold (45dBZ) is reached at a height exceeding the -5°C level. The intended primary seeding system for NHRE is the airborne, vertically launched cloud-seeking rockets. However, it was not used during summer of 1972 due to unanticipated developmental problems. Hence, cloud-base seeding was tried using end-burning flares carried under the fuselage of the aircraft. The summer of 1972 was the first operational program of the NHRE suppression experiment. The second season of field operations during the summer of 1973 provided a rather disappointingly low number of hail days (6) for the ongoing hail suppression test. Thus the results to prove the feasibility of hail suppression are as yet not available.

The following preliminary results of updraft velocities and hailstone growth trajectories are tentative at best (NHRE project plan 1975-1980):

i) Updraft velocities at cloud base are in the range of 5 to 10 $msec^{-1}$, and 15 to 25 $msec^{-1}$ at the core.

ii) The environment of graupel embryo formation in NE Colorado storms is very probably of low liquid water content.

iii) It does not appear that high water contents are important for hail formation.

iv) Deuterium content profiles in large hailstones have shown a simple up-and-down motion without any recirculation.

v) The reflectivity maxima within the overhang of a large hailstorm indicated the absence of recirculating trajectories.

These results, though tentative, do support the author's model of hailstone growth as discussed in Chapter 8.

After reading the project plans and annual reports of NHRE over the period 1972-75, it seems the project is rather unproductive to date as it is experiencing a number of problems related to changes in scientific strategy, management and staff, and questionable and an ill-defined starting hypothesis for investigations. In addition, there has been a rather small amount of microphysical work done related to hail suppression problem, inadequate synthesis and integration of activities, and considerable backlog in the reduction and analysis of radar data and hail/rain data. The most important handicap is the comparative lack of well-defined, isolated storms and relatively small sample of them in the target area to give any conclusive, statistically significant results in a five-year period. Further complication is due to the dominance of the statistical data by the few <u>severe</u> storms.

APPENDIX D

USEFUL PHYSICAL CONSTANTS

Avogadro's constant = 6.022×10^{23} mol^{-1}

Universal gas constant = 8.314×10^{7} erg mol^{-1} K^{-1}

Boltzmann's constant = 1.381×10^{-16} erg K^{-1}

Molecular weight of dry air = 28.966

Molecular weight of water vapor = 18.015

Specific heat capacity of dry air:

constant pressure, 0.240 cal g^{-1} K^{-1}

constant volume, 0.171 cal g^{-1} K^{-1}

Density of air at 0C and 1000 mb pressure

= 1.276×10^{-3} g cm^{-3}

(for other temperatures and pressures multiply by 0.273 p/T)

Dynamic viscosity of air = 1.718×10^{-14} poise at 0C

Kinematic viscosity of air = 0.135 cm^{2} sec^{-1} at 0C

Diffusion coefficient of water vapor in air

= 5.80×10^{5} cal cm^{-1} sec^{-1} K^{-1} at 0C

Heats of phase transformations of waters:

Vaporization	586.0 cal g^{-1} at 20C
	597.3 cal g^{-1} at 0C
	608.9 cal g^{-1} at -20C
Fusion	79.7 cal g^{-1} at 0C
	69.0 cal g^{-1} at -20C

Sublimation 677.0 cal g^{-1} at 0C

677.9 cal g^{-1} at -20C

Specific heat of liquid water = 1.007 cal g^{-1} K^{-1} at 0C

Specific heat of ice = 0.503 cal g^{-1} K^{-1} at 0C

Density of saturated water vapor

over water, 3.93 gm^{-3} at 270K

1.87 gm^{-3} at 260K

over ice, 3.82 gm^{-3} at 270K

1.65 gm^{-3} at 260K

Acceleration of gravity

Standard value = 980.665 cm sec^{-2} at sea level

(It decreases with height about 0.3 cm sec^{-2} per kilometer in free atmosphere.)

The standard atmosphere (Lower Levels)

(from the Smithsonian Meteorological Tables, 1951)

Altitude km	Temperature °C	Pressure mb	Altitude km	Temperature °C	Pressure mb
0	15	1013.25	10.769	-55	234.53
2	2	794.90	12	-55	193.38
4	-11	616.29	14	-55	141.35
6	-24	471.65	16	-55	103.30
8	-37	355.82	18	-55	75.53
10	-50	264.19	20	-55	55.21

Saturation vapor pressures over pure liquid water (e_w) and over pure ice (e_i) as functions of temperature

T (°C)	e_w (mb)	e_i (mb)	T	e_w	e_i	T	e_w	T	e_w
-40	0.1891	o.1283	-20	1.2538	1.032	1	6.565	22	26.430
-39	0.2097	0.1436	-19	1.3661	1.135	2	7.054	24	29.831
-38	0.2322	0.1606	-18	1.4874	1.248	3	7.574	26	33.608
-37	0.2570	0.1794	-17	1.6183	1.371	4	8.128	28	37.793
-36	0.2841	0.2002	-16	1.7594	1.505	5	8.718	30	42.427
-35	0.3138	0.2232	-15	1.9114	1.651	6	9.345	31	44.924
-34	0.3463	0.2487	-14	2.0751	1.810	7	10.012	32	47.548
-33	0.3817	0.2768	-13	2.2512	1.983	8	10.720	33	50.303
-32	0.4204	0.3078	-12	2.4405	2.171	9	11.473	34	53.197
-31	0.4627	0.3420	-11	2.6438	2.375	10	12.271	35	56.233
-30	0.5087	0.3797	-10	2.8622	2.597	11	13.118	36	59.418
-29	0.5588	0.4212	- 9	3.0965	2.837	12	14.016	37	62.759
-28	0.6133	0.4668	- 8	3.3478	3.097	13	14.967	38	66.260
-27	0.6726	0.5169	- 7	3.6171	3.397	14	15.975	39	69.930
-26	0.7369	0.5719	- 6	3.9055	3.684	15	17.042	40	73.773
-25	0.8068	0.6322	- 5	4.2142	4.014	16	18.171	41	77.798
-24	0.8826	0.6983	- 4	4.5444	4.371	17	19.365	42	82.011
-23	0.9647	0.7708	- 3	4.8974	4.756	18	20.628	43	86.419
-22	1.0536	0.8501	- 2	5.2745	5.173	19	21.962	44	91.029
-21	1.1498	0.9366	- 1	5.6772	5.622	20	23.371	45	95.850
			0	6.1070	6.106				

Terminal velocities of water drops in still air, pressure 760 mm, temperature 20°C

Gunn and Kinzer (1949)

Drop diameter (mm)	Terminal velocity (m sec^{-1})	Mass (μg)
0.1	0.27	0.524
0.2	0.72	4.19
0.3	1.17	14.14
0.4	1.62	33.5
0.5	2.06	65.5
0.6	2.47	113.1
0.7	2.87	179.6
0.8	3.27	268
0.9	3.67	382
1.0	4.03	524
1.2	4.64	905
1.4	5.17	1,437
1.6	5.65	2,140
1.8	6.09	3,050
2.0	6.49	4,190
2.2	6.90	5,580
2.4	7.27	7,240
2.6	7.57	9,200
2.8	7.82	11,490
3.0	8.06	14,140
3.2	8.26	17,160
3.4	8.44	20,600
3.6	8.60	24,400
3.8	8.72	28,700
4.0	8.83	33,500
4.2	8.92	38,800
4.4	8.98	44,600
4.6	9.03	51,000
4.8	9.07	57,900
5.0	9.09	65,500
5.2	9.12	73,600
5.4	9.14	82,400
5.6	9.16	92,000
5.8	9.17	102,200

REFERENCES AND AUTHOR INDEX

AMERICAN METEOROLOGICAL SOCIETY (1959) Glossary of Meteorology (ed. R. E. Huschke), p. 268, 269, 297, 520. American Meteorological Society, Boston. [Chapter 4]

APPLEMAN, H. S. (1959) An investigation into the formation of hail. Nubila 2, 28. [Chapter 2, 10, 11]

ARENBERG, D. L. (1938) The formation of irregularly shaped hailstones. Mon. Wea. Rev., 66, 274. [Chapter 4]

ARNOLD, J. E. (1961) Some characteristics of severe Texas thunderstorms. Utilization of AN/CPS-9 Radar in Weather Analysis and Forecasting, p. 47. Final Report - Contract AF 19 (604)-6136, A. and M. College of Texas [Now Texas A&M University]. [Chapter 6]

ATLAS, D. (1963) Radar analysis of severe storms. Met. Monogr. 5, No. 27, p. 177. American Meteorological Society, Boston, [Chapter 6, 10]

__________ (1964) Advances in radar meteorology. Advances in Geophysics, 10, p. 317. Academic Press, New York. [Chapter 6, 10]

__________ BOOKER, D. R., BYERS, H., DOUGLAS, R. H., FUJITA, T., HOUSE, D. C., LUDLAM, F. H., MALKUS, J. S., NEWTON, C. W., OGURA, Y., SCHLEUSENER, R. A., VONNEGUT, B., AND WILLIAMS, R. T. (1963) Severe Local Storms. Met. Monogr.

5, No. 27. American Meteorological Society, Boston. [Chapter 7]

__________ and LUDLAM, F. H. (1960) Multi-wavelength reflectivity of hailstorms. Imperial College Dept. Met. Tech. Note No. 9, Contract AF 61 (052)-254, London. [Chapter 2, 6, Appendix B]

__________ __________ (1961) Multi-wavelength radar reflectivity of hailstorms. Quart. J. Roy. Met. Soc. 87, 523. [Chapter 6]

__________ and WEXLER, R. (1963) Back-scatter by oblate ice spheroids. J. Atmos. Sci. 20, 48. [Chapter 6]

AUER, A. H. (1972) Distribution of graupel and hail with size. Mon. Wea. Rev. 100, No. 5, 325. [Chapter 2]

AUFDERMAUR, A. N., LIST, R., MAYES, W. C., and DE QUERVAIN, M. R. (1963) Kristallachsenlagen in Hagelkörnern. Z. angew. Math. Phys. (ZAMP) 14, 574. [Chapter 5]

AUFM KAMPE, H. J. and WEICKMANN, H. R. (1957) Physics of clouds. Met. Monogr. 3, No. 18, p. 182. American Meteorological Scoiety, Boston. [Chapter 7]

BAILEY, I. H. and MACKLIN, W. C. (1968) The surface configuration and internal structure of artificial hailstones. Quart. J. Roy. Met. Soc. 94, 1. [Chapter 4, 5, Appendix B]

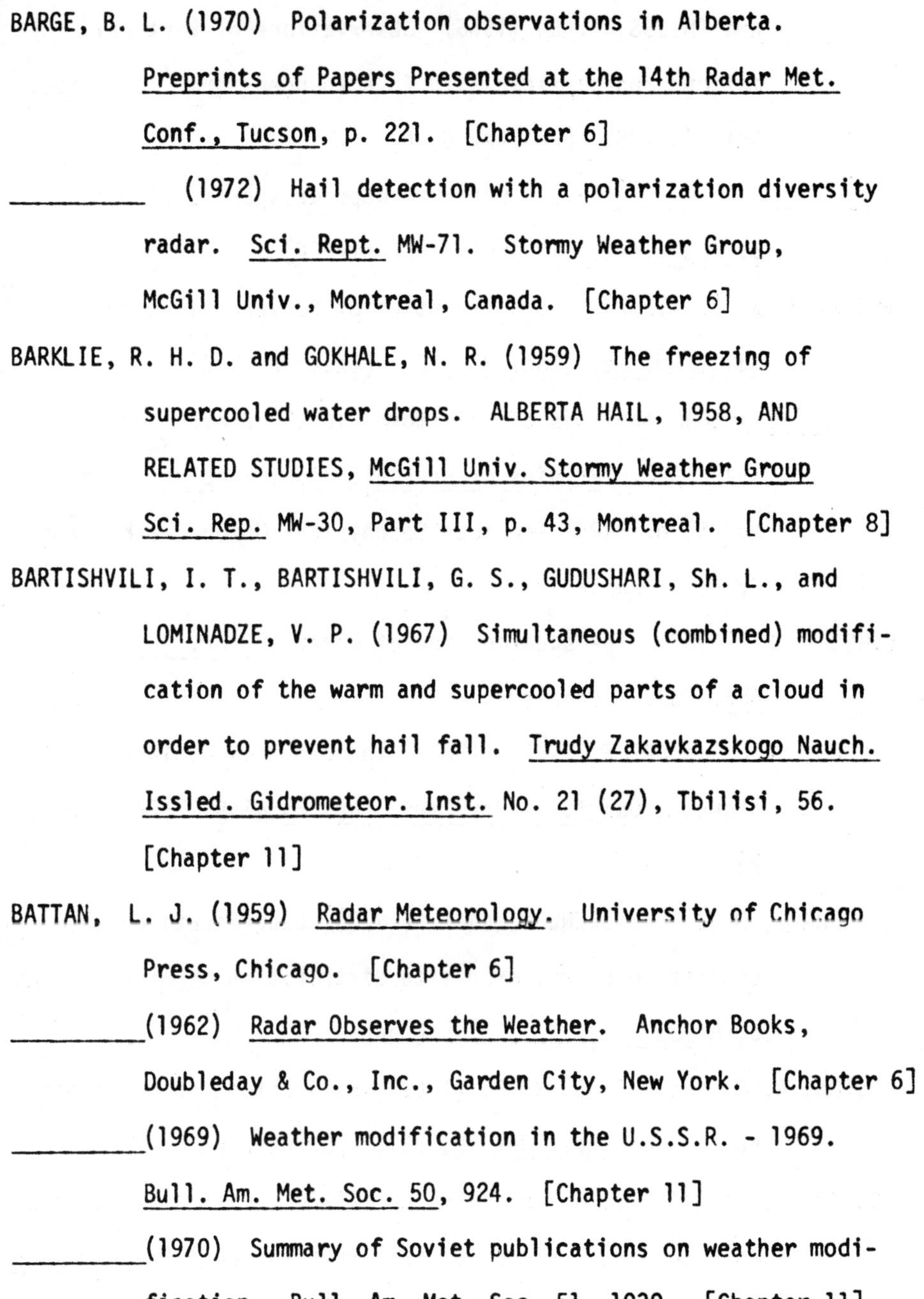

BARGE, B. L. (1970) Polarization observations in Alberta. Preprints of Papers Presented at the 14th Radar Met. Conf., Tucson, p. 221. [Chapter 6]

__________ (1972) Hail detection with a polarization diversity radar. Sci. Rept. MW-71. Stormy Weather Group, McGill Univ., Montreal, Canada. [Chapter 6]

BARKLIE, R. H. D. and GOKHALE, N. R. (1959) The freezing of supercooled water drops. ALBERTA HAIL, 1958, AND RELATED STUDIES, McGill Univ. Stormy Weather Group Sci. Rep. MW-30, Part III, p. 43, Montreal. [Chapter 8]

BARTISHVILI, I. T., BARTISHVILI, G. S., GUDUSHARI, Sh. L., and LOMINADZE, V. P. (1967) Simultaneous (combined) modification of the warm and supercooled parts of a cloud in order to prevent hail fall. Trudy Zakavkazskogo Nauch. Issled. Gidrometeor. Inst. No. 21 (27), Tbilisi, 56. [Chapter 11]

BATTAN, L. J. (1959) Radar Meteorology. University of Chicago Press, Chicago. [Chapter 6]

__________ (1962) Radar Observes the Weather. Anchor Books, Doubleday & Co., Inc., Garden City, New York. [Chapter 6]

__________ (1969) Weather modification in the U.S.S.R. - 1969. Bull. Am. Met. Soc. 50, 924. [Chapter 11]

__________ (1970) Summary of Soviet publications on weather modification. Bull. Am. Met. Soc. 51, 1030. [Chapter 11]

_________ and THEISS, J. B. (1966) Observations of vertical motions and particle size in a thunderstorm. J. Atmos. Sci. 23, 78. [Chapter 6, 8]

_________ _________ (1972) Observed Doppler spectra of hail. J. Appl. Met. 11, 1001. [Chapter 6]

BECKWITH, W. B. (1957) Characteristics of Denver hailstorms. Bull. Am. Met. Soc. 38, 20. [Chapter 2, 3]

_________ (1960) Analysis of hailstorms in the Denver network, 1949-1958. Physics of Precipitation, Geophys. Monogr. No. 5, p. 348. American Geophysical Union, Washington, D. C. [Chapter 2, 3]

BIGG, E. K. (1953) The supercooling of water. Proc. Phys. Soc. B66, 688. [Chapter 8]

BILHAM, E. G. and RELF, E. F. (1937) The dynamics of large hailstones. Quart. J. Roy. Met. Soc. 63, 149. [Chapter 4, 7]

BLANCHARD, D. C. and SPENCER, A. T. (1970) Experiments on the generation of raindrop size distribution by drop breakup. J. Atmos. Sci. 27, 101. [Chapter 8]

BOSTON, R. C. and ROGERS, R. R. (1969) Hail detection by Doppler radar. J. Appl. Met. 8, 837. [Chapter 6]

BOUTIN, C., ISAKA, H., and SOULAGE, G. (1970) Statistical studies on French operations for hail suppression. Preprints of Papers Presented at the 2nd National Conf. on Weather

Modification, Santa Barbara, p. 134. [Chapter 11]

BRIGGS, G. A. (1968) Hailstones, starfish, and daggers - spiked hail falls in Oak Ridge, Tennessee. Mon. Wea. Rev. 96, 744. [Chapter 4]

BROOKS, C. F. (1922) The local, or heat, thunderstorm. Mon. Wea. Rev. 50, 281. [Chapter 7]

BROWNING, K. A. (1963) The growth of large hail within a steady updraught. Quart. J. Roy. Met. Soc. 89, 490. [Chapter 8]

__________(1965) A family outbreak of severe local storms - a comprehensive study of the storms in Oklahoma on 26 May 1963 - Part I. Air Force Cambridge Research Laboratories - 65-695(1) Special Report No. 32, Bedford, Massachusetts. [Chapter 6]

__________(1966) The lobe structure of giant hailstones. Quart. J. Roy. Met. Soc. 92, 1. [Chapter 5]

__________ and BEIMERS, J. G. D. (1967) The oblateness of large hailstones. J. Appl. Met. 6, 1075. [Chapter 4]

__________DONALDSON, R. J., JR., and LAMKIN, W. E. (1963) Severe local storms near Oklahoma City, 26 May 1963. Conference Review, 3rd Conf. on Severe Local Storms, Urbana, Illinois. [Chapter 6]

__________HALLETT, J., HARROLD, T. W., and JOHNSON, D. (1968) The collection and analysis of freshly fallen hailstones.

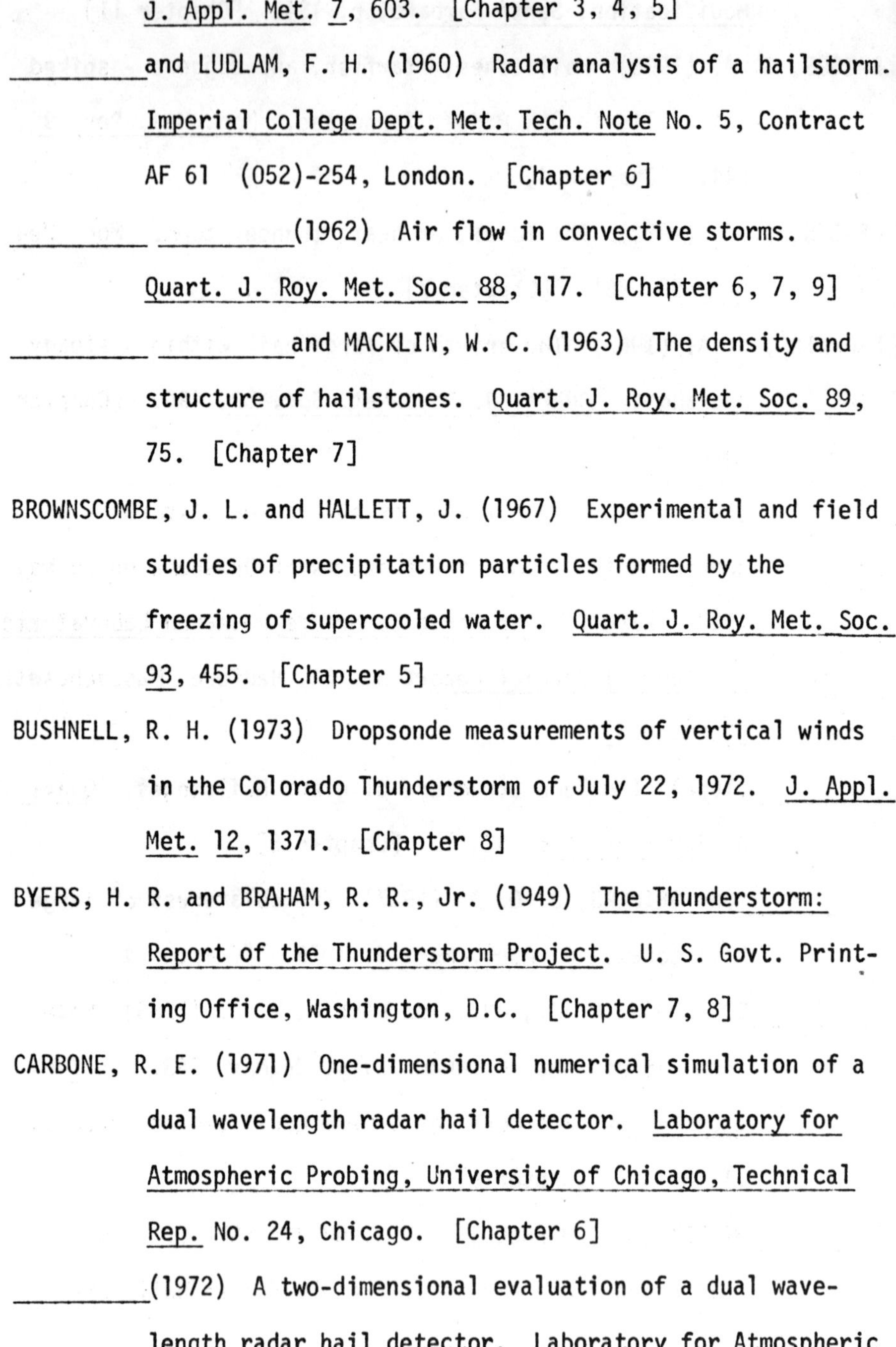

J. Appl. Met. 7, 603. [Chapter 3, 4, 5]

__________ and LUDLAM, F. H. (1960) Radar analysis of a hailstorm. Imperial College Dept. Met. Tech. Note No. 5, Contract AF 61 (052)-254, London. [Chapter 6]

__________ __________ (1962) Air flow in convective storms. Quart. J. Roy. Met. Soc. 88, 117. [Chapter 6, 7, 9]

__________ __________ and MACKLIN, W. C. (1963) The density and structure of hailstones. Quart. J. Roy. Met. Soc. 89, 75. [Chapter 7]

BROWNSCOMBE, J. L. and HALLETT, J. (1967) Experimental and field studies of precipitation particles formed by the freezing of supercooled water. Quart. J. Roy. Met. Soc. 93, 455. [Chapter 5]

BUSHNELL, R. H. (1973) Dropsonde measurements of vertical winds in the Colorado Thunderstorm of July 22, 1972. J. Appl. Met. 12, 1371. [Chapter 8]

BYERS, H. R. and BRAHAM, R. R., Jr. (1949) The Thunderstorm: Report of the Thunderstorm Project. U. S. Govt. Printing Office, Washington, D.C. [Chapter 7, 8]

CARBONE, R. E. (1971) One-dimensional numerical simulation of a dual wavelength radar hail detector. Laboratory for Atmospheric Probing, University of Chicago, Technical Rep. No. 24, Chicago. [Chapter 6]

__________ (1972) A two-dimensional evaluation of a dual wavelength radar hail detector. Laboratory for Atmospheric

Probing, University of Chicago, Technical Rep. No. 25, Chicago. [Chapter 6]

CARRAS, J. N. and MACKLIN, W. C. (1973) The shedding of accreted water during hailstone growth. Quart. J. Roy. Met. Soc. 99, 639. [Appendix A]

__________ __________ (1975) Air bubbles in accreted ice. Quart. J. Roy. Met. Soc. 101, 127. [Chapter 5]

CARTE, A. E. (1961) Air bubbles in ice. Proc. Phys. Soc. 77, 757. [Chapter 5]

__________ (1963) Some characteristics of Alberta hailstorms. ALBERTA HAIL STUDIES 1962/63, McGill Univ. Stormy Weather Group Sci. Rep. MW-36, p. 1, Montreal. [Chapter 4]

__________ (1964) Hailstorms in Johannesburg, Pretoria and surroundings on January 15 and 16, 1964. CSIR Research Rep. No. 228, Pretoria, South Africa. [Chapter 2, 3, 4]

__________ (1968) Mine shafts as a cloud physics facility. Proc. Int. Conf. on Cloud Physics, Toronto, p. 384. [Chapter 4]

__________ and BASSON, I. L. (1970) Hail in the Pretoria-Witwatersrand area 1962-1969. CSIR Research Rep. No. 293, Pretoria, South Africa. [Chapter 2, 3, 4]

__________ DOUGLAS, R. H., EAST, C., GUNN, K. L. S., HITSCHFELD, W., MARSHALL, J. S., and STANSBURY, E. J. (1961) Alberta Hail Studies 1961. McGill Univ. Stormy Weather

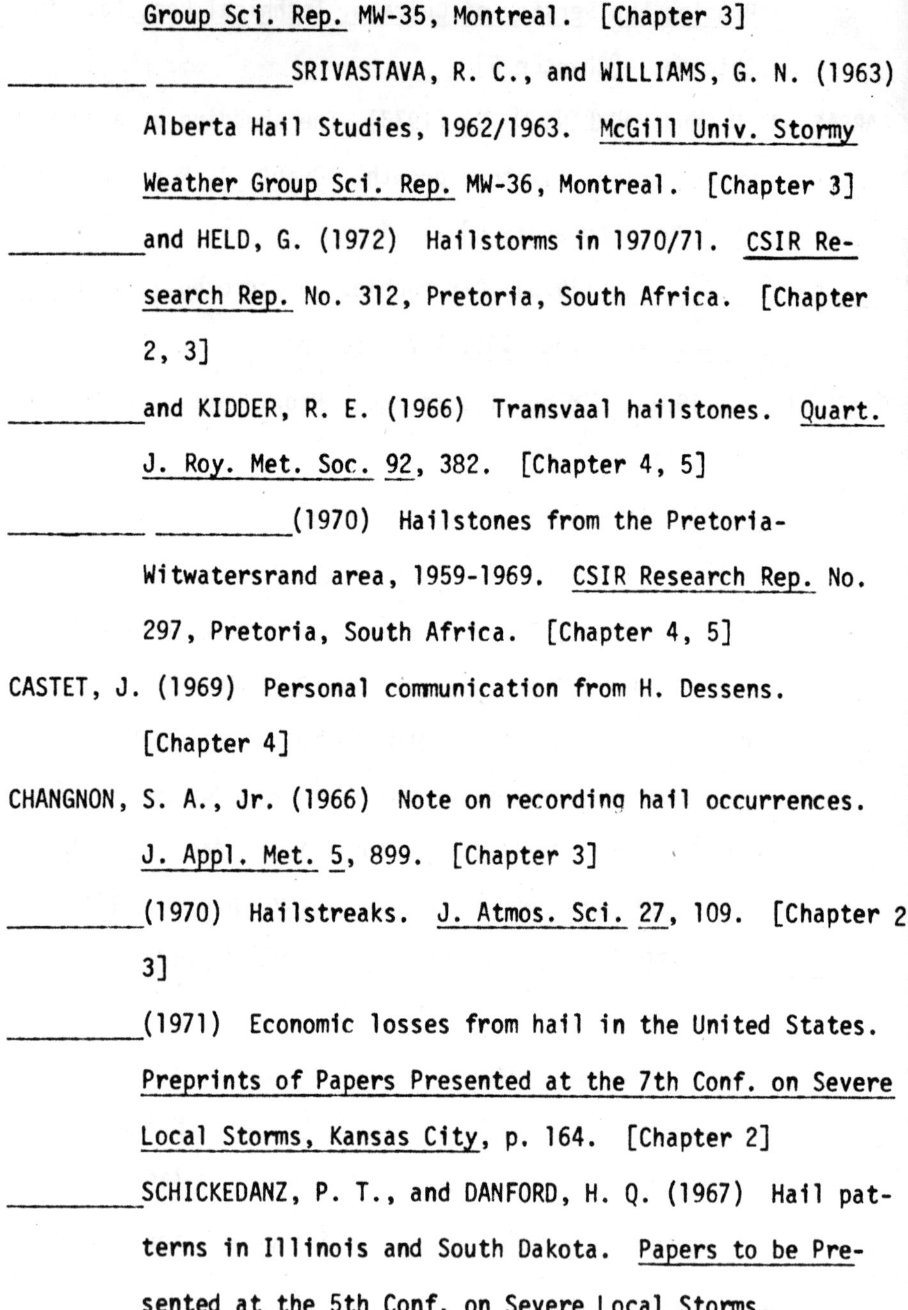

Group Sci. Rep. MW-35, Montreal. [Chapter 3]

—————— ——————SRIVASTAVA, R. C., and WILLIAMS, G. N. (1963) Alberta Hail Studies, 1962/1963. *McGill Univ. Stormy Weather Group Sci. Rep.* MW-36, Montreal. [Chapter 3]

——————and HELD, G. (1972) Hailstorms in 1970/71. *CSIR Research Rep.* No. 312, Pretoria, South Africa. [Chapter 2, 3]

——————and KIDDER, R. E. (1966) Transvaal hailstones. *Quart. J. Roy. Met. Soc.* *92*, 382. [Chapter 4, 5]

—————— ——————(1970) Hailstones from the Pretoria-Witwatersrand area, 1959-1969. *CSIR Research Rep.* No. 297, Pretoria, South Africa. [Chapter 4, 5]

CASTET, J. (1969) Personal communication from H. Dessens. [Chapter 4]

CHANGNON, S. A., Jr. (1966) Note on recording hail occurrences. *J. Appl. Met.* *5*, 899. [Chapter 3]

——————(1970) Hailstreaks. *J. Atmos. Sci.* *27*, 109. [Chapter 2, 3]

——————(1971) Economic losses from hail in the United States. *Preprints of Papers Presented at the 7th Conf. on Severe Local Storms, Kansas City*, p. 164. [Chapter 2]

——————SCHICKEDANZ, P. T., and DANFORD, H. Q. (1967) Hail patterns in Illinois and South Dakota. *Papers to be Presented at the 5th Conf. on Severe Local Storms,*

St. Louis, p. 325. [Chapter 3]

_________ and STOUT, G. E. (1964) A detailed investigation of an Illinois hailstorm on August 8, 1963. Crop-Hail Insurance Actuarial Association Rep. No. 18, Chicago, Illinois. [Chapter 6]

CHARLTON, R. B. and LIST, R. (1972) Hail size distributions and accumulation zones. J. Atmos. Sci. 29, 1182. [Chapter 7]

CHISHOLM, A. J. (1968) Observations by 10-cm radar of an Alberta hailstorm in a sheared environment. Proc. 13th Radar Met. Conf., Montreal, p. 82. [Chapter 6]

_________ and ENGLISH, M. (1973) Alberta Hailstorms. Met. Monographs, vol. 14, No. 36. [Chapter 3, 7]

DANIELSEN, F. D., BLECK, R. and MORRIS, D. A. (1972) Hail growth by stochastic collection in a cumulus model. J. Atmos. Sci. 29, 135. [Chapter 7]

DAVIES-JONES, R. P. (1974) Discussion of measurements inside high-speed thunderstorm updrafts. J. Appl. Met. 13, 710. [Chapter 8]

DAVIES, C. N. (1939) Unpublished Report, Ministry of Supply, United Kingdom, London. [Chapter 8]

DAVIS, W. M. (1894) Elementary Meteorology. Ginn & Co., Boston. [Chapter 7]

DEIBERT, R. J. and RENICK, J. H. (1975) Alberta hail project field program 1974. Int. Wea. Mod. Board, Three Hills, Alberta, Canada. [Chapter 3]

DENNIS, A. S. and MUSIL, D. J. (1973) Calculations of hailstone growth and trajectories in a simple cloud model. J. Atmos. Sci. 30, 278. [Chapter 7]

__________ SCHOCK, C. A., and KOSCIELSKI, A. (1970) Characteristics of hailstorms of Western South Dakota. J. Appl. Met. 9, 127. [Chapter 6, 8]

DE QUERVAIN, M. R. (1954) Die Metamorphose des Schneekristalls. Verhandlungen der Schweiz. Naturforschenden Gesellschaft, Davos, 114. [Chapter 5]

DESSENS, H. (1960) Severe hailstorms are associated with very strong winds between 6,000 and 12,000 meters. Physics of Precipitation, Geophys. Monogr. No. 5, p. 333. American Geophysical Union, Washington, D.C. [Chapter 7, 8]

__________ et al. (1967) Association D'Etudes des Moyens de Luttes contre les Fleaux Atmospheriques. No. 15, Toulouse, France. [Chapter 11]

DESSENS, J. (1968) Experience de suppression de la grele dans le sud-ouest de la France. Proc. Int. Conf. on Cloud Physics, Toronto, p. 773. [Chapter 11]

__________ and LACAUX, J. (1972) Ground seeding for hail prevention in South-Western France: Possible overstepping of an economical efficiency level from 1963. Preprints - 3rd Conf. on Weather Modification of the Am. Met. Soc., Rapid City, p. 268. [Chapter 11]

DONALDSON, R. J., Jr. (1958) Vertical profiles of radar reflectivity in thunderstorms. Proc. 7th Weather Radar Conf., Miami, p. 88. [Chapter 6]

__________(1961) Radar reflectivity profiles in thunderstorms. J. Met. 18, 292. [Chapter 6, 8, 10]

__________(1962) Radar observations of a tornado thunderstorm in vertical section. National Severe Storms Project Rep. No. 8 (preprint). U. S. Weather Bureau, Washington, D.C. [Chapter 6, 7]

__________CHMELA, A. C., and SHACKFORD, C. R. (1960) Some behavior patterns of New England hailstorms. Physics of Precipitation, Geophys. Monogr. No. 5, p. 354. American Geophysical Union, Washington, D.C. [Chapter 2, 3, 4, 8]

__________and WEXLER, R. (1969) Flight hazards in thunderstorms determined by Doppler velocity variance. J. Appl. Met. 8, 128. [Chapter 6]

DORSEY, N. E. (1940) Properties of ordinary water substance. Am. Chem. Soc. Month. 81. [Chapter 5]

DOUGLAS, R. H. (1960) Size distributions, ice contents, and radar reflectivities of hail in Alberta. Nubila 3, 5.

[Chapter 2]

________(1963) Recent hail research: A review. Met. Monogr. 5, No. 27, p. 157. American Meteorological Society, Boston. [Chapter 2, 4, 6]

________and HITSCHFELD, W. F. (1958) Studies of Alberta hailstorms 1957. McGill Univ. Stormy Weather Group Sci. Rep. MW-27, Montreal. [Chapter 6]

________ ________(1959) Patterns of hailstorms in Alberta. Quart. J. Roy. Met. Soc. 85, 105. [Chapter 3, 10]

________ ________(1961) Radar reflectivities of hail samples Proc. 9th Weather Radar Conf., Kansas City, p. 147. [Chapter 6]

________MARSHALL, J. S., and BARKLIE, R. H. D. (1962) Interim account of hail studies - November 1960. McGill Univ. Stormy Weather Group Sci. Rep. MW-34, Montreal. (Reprint of 1960 report) [Chapter 6]

DRAKE, J. C. and MASON, B. J. (1966) The melting of small ice spheres and cones. Quart. J. Roy. Met. Soc. 92, 500. [Appendix B]

ECCLES, P. J. and ATLAS, D. (1973) A dual-wavelength radar hail detector. J. Appl. Met. 12, 847. [Chapter 6]

FACY, L., MERLIVAT, L., NIEF, G., and ROTH, E. (1963) The study of the formation of hailstones by isotopic analysis. J. Geophys. Res. 68, 3841. [Chapter 9]

FANKHAUSER, J. C. (1971) Thunderstorm-environment interactions determined from aircraft and radar observations. Mon. Wea. Rev. 99, 171. [Chapter 8]

FAVREAU, R. F. and GOYER, G. G. (1967) The effect of shock waves on a hailstone model. J. Appl. Met. 6, 326. [Chapter 11]

FAWBUSH, E. J. and MILLER, R. C. (1953) A method for forecasting hailstone size at the earth's surface. Bull. Am. Met. Soc. 34, 235. [Chapter 10]

FEDERER, B. and WALDVOGEL (1975) Hail and raindrop size distributions from a Swiss multicell storm. J. Appl. Met. 14, 91. [Chapter 2, 4]

FINDEISEN, W. (1940) Über die Entstehung der Gewitterelektrizität. Met. Z. 57, 201. [Chapter 7]

FLETCHER, N. H. (1962) The Physics of Rainclouds, p. 126. Cambridge University Press, London. [Chapter 8]

FLORA, S. D. (1956) Hailstorms of the United States. University of Oklahoma Press, Norman. [Chapter 2]

FOOTE, G. B. and DU TOIT, P. S. (1969) Terminal velocity of raindrops aloft. J. Appl. Met. 8, 249. [Chapter 8]

FOSTER, D. S. and BATES, F. C. (1956) A hail size forecasting technique. Bull. Am. Met. Soc. 37, 135. [Chapter 10]

FRANK, S. R. (1957) Survey and history of hail-suppression operations in the United States. Final Report of the

Advisory Committee on Weather Control, Volume II, p. 264 U. S. Govt. Printing Office, Washington, D.C. [Chapter 11]

FRIEDMAN, I., REDFIELD, A. C., SCHOEN, B., and HARRIS, J. (1964) The variation of the deuterium content of natural waters in the hydrological cycle. Revs. Geophys. 2, 177. [Chapter 9]

FRISBY, E. M. and SANSOM, H. W. (1967) Hail incidence in the tropics. J. Appl. Met. 6, 339. [Chapter 2]

FUJITA, T. and BYERS, H. R. (1960) Model of a hail cloud as revealed by photogrammetric analysis. Univ. of Chicago Tech. Rep. No. 3, Chicago. [Chapter 8]

GEOTIS, S. G. (1963) Some radar measurements of hailstorms. J. Appl. Met. 2, 270. [Chapter 6, 7, 10]

GITLIN, S. N., GOYER, G., and HENDERSON, T. J. (1968) The liquid water content of hailstones. J. Atmos. Sci. 25, 97. [Chapter 4, 5 and 7]

GOKHALE, N. R. (1965a) Dependence of freezing temperature of supercooled water drops on rate of cooling. J. Atmos. Sci. 22, 212. [Chapter 8]

__________ (1965b) Comparison of ice nucleating efficiencies of chemical aerosols in a supercooled cloud and in bulk water. Proc. Int. Conf. on Cloud Physics, Tokyo and

Sapporo, p. 176. [Chapter 11]

________and GOOLD, J., Jr., (1968) Droplet freezing by surface nucleation. J. Appl. Met. 7, 870. [Chapter 11]

________and LEWINTER, O. (1971) Microcinematographic studies of contact nucleation. J. Appl. Met. 10, 469. [Chapter 11]

________and RAO, K. M. (1969a) Theory of hail growth. J. Rech. Atmos. 4, 153. [Chapter 4, 8, 10]

________ ________(1969b) Accumulation of large hydrometeors in the upper part of an intensive updraft in a cumulonimbus. Preprints of Papers Presented at the 6th Conf. on Severe Local Storms, Chicago, p. 59. (Unpublished manuscript) [Chapter 8]

________ ________(1970) Estimation of the occurrence of hail and hailstone sizes. Preprints of Papers Presented at the Conference on Cloud Physics, Fort Collins, Colorado (Am. Met. Soc.), p. 59. [Chapter 10]

________ ________(1971) Comments on 'The Analysis of a Hailstone' by W. C. Macklin, L. Merlivat and C. M. Stevenson. Quart. J. Roy. Met. Soc. 97, 575. [Chapter 9]

________ ________(1972) Some new suggestions to identify thunderstorms with hail. Journal de Recherches Atmospheriques - Vol. VI, No. 1-2-3, p. 187. [Chapter 10]

________and SPENGLER, J. D. (1972) Freezing of freely suspended, supercooled water drops by contact nucleation. J. Appl. Met. 11, 157. [Chapter 11]

__________ __________(1973) Artificial growth of "icicle" lobe structure of a hailstone in a wind tunnel. J. Appl. Met. 12, 418. [Chapter 5]

GORI, E. G., MUSSO, G., and PAPEE, H. M. (1971) On the hailstorms of Asti Province. Part II: Past hail-suppression endeavours. Geofisica e Meteorologia XX, p. 24. [Chapter 11]

GOYER, G. G. (1971) [Editor] The Joint Hail Research Project - Summer 1970, Summary Report. National Center for Atmospheric Research, Boulder, Colorado. [Chapter 10]

__________LIN, S. S., GITLIN, S. N. and PLOOSTER, M. N. (1968) The rate of heat transfer to dry and wet hailstone surfaces and its application to the freezing of spongy hail in a cloud model. Proc. Inter. Conf. Cloud Phys., Toronto, Canada, August 26-30, p. 427. [Chapter 5]

__________and ROADS, J. O. (1971) The mapping of hailswaths by airborne infrared radiometry. Proc. Int. Conf. on Weather Modification, Canberra, p. 231. [Chapter 3]

__________and WOOD, J. M. (1972) The radar climatology of thunderstorms in Northeastern Colorado. Preprints - 3rd Conf. on Weather Modification of the Am. Met. Soc., Rapid City, p. 248. [Chapter 2]

GRANDOSO, H. N. (1966) Distribución Temporal y Geográfica del Granizo en la Provincia de Mendoza y su relacion con algunos Parametros Meteorológicos. Universidad de Buenos Aires, Facultad de Ciencias Exactas y Naturales, Contribuciones Centrificas, Serie Meteorologia, Volumen 1, Numero 7. [Chapter 2]

________ and IRIBARNE, J. V. (1961) The experiment on hail prevention in Mendoza, Argentina. Proc. Int. Conf. on Cloud Physics, Canberra, p. 64. [Chapter 11]

________ ________ (1963) Evaluation of the first three years in a hail prevention experiment in Mendoza (Argentina). Z. Angew. Math. Phys. (ZAMP) 14, 549. [Chapter 11]

GRIMMINGER, G. (1933) The upward speed of an air current necessary to sustain a hailstone. Mon. Wea. Rev. 61, 198. [Chapter 10]

GUNN, R. and KINZER, G. D. (1949) The terminal velocity of fall for water droplets in stagnant air. J. Met. 6, 243. [Chapter 8, Appendix D]

GUSEV, M. A. (1966) Errors resulting in the shelling of hail zones with "Elbrus-II" non-fragmenting shells using the KS-19 artillery gun. Trudy Vysok. Geofiz. Inst. No. 3 (5), Leningrad, 227. [Chapter 11]

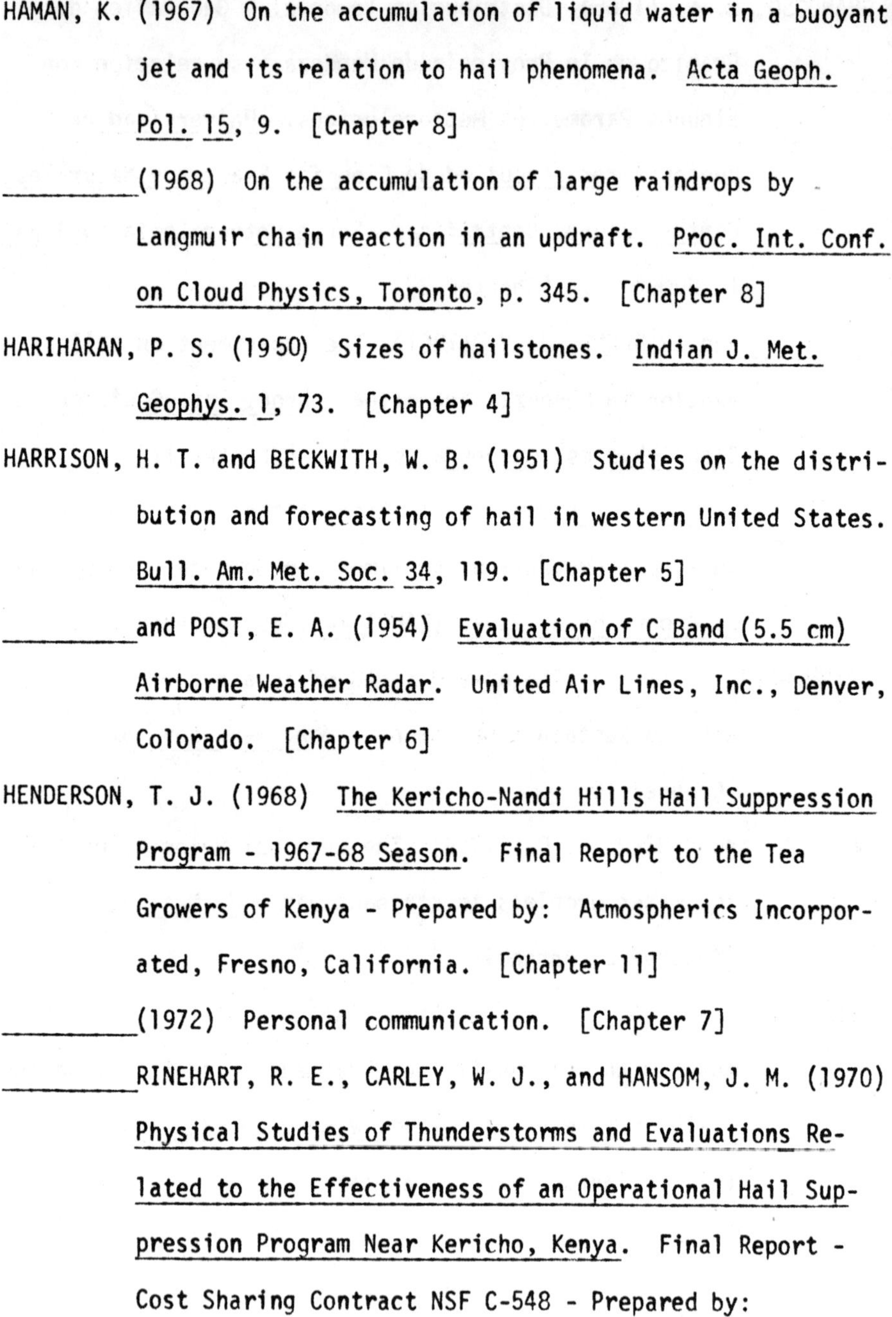

HAMAN, K. (1967) On the accumulation of liquid water in a buoyant jet and its relation to hail phenomena. Acta Geoph. Pol. 15, 9. [Chapter 8]

__________(1968) On the accumulation of large raindrops by Langmuir chain reaction in an updraft. Proc. Int. Conf. on Cloud Physics, Toronto, p. 345. [Chapter 8]

HARIHARAN, P. S. (1950) Sizes of hailstones. Indian J. Met. Geophys. 1, 73. [Chapter 4]

HARRISON, H. T. and BECKWITH, W. B. (1951) Studies on the distribution and forecasting of hail in western United States. Bull. Am. Met. Soc. 34, 119. [Chapter 5]

__________and POST, E. A. (1954) Evaluation of C Band (5.5 cm) Airborne Weather Radar. United Air Lines, Inc., Denver, Colorado. [Chapter 6]

HENDERSON, T. J. (1968) The Kericho-Nandi Hills Hail Suppression Program - 1967-68 Season. Final Report to the Tea Growers of Kenya - Prepared by: Atmospherics Incorporated, Fresno, California. [Chapter 11]

__________(1972) Personal communication. [Chapter 7]

__________RINEHART, R. E., CARLEY, W. J., and HANSOM, J. M. (1970) Physical Studies of Thunderstorms and Evaluations Related to the Effectiveness of an Operational Hail Suppression Program Near Kericho, Kenya. Final Report - Cost Sharing Contract NSF C-548 - Prepared by:

Atmospherics Incorporated, Fresno, California. [Chapter 2, 11]

HIGUCHI, K. (1958) The etching of ice crystals. Acta Met. 6, 636. [Chapter 5]

HITSCHFELD, W. F. (1960) The motion and erosion of convective storms in severe vertical wind shear. J. Met. 17, 270. [Chapter 8]

__________(1971) [Editor] Hail research at McGill, 1956-1971. McGill Univ. Stormy Weather Group Sci. Rep. MW-68, Montreal. [Chapter 6]

__________and DOUGLAS, R. H. (1963) A theory of hail growth based on studies of Alberta storms. Z. angew. Math. Phys. (ZAMP) 14, 554. [Chapter 7, 8]

HOFFER, T. E. (1961) A laboratory investigation of droplet freezing. J. Met. 18, 766. [Chapter 11]

HOLROYD, E. W. (1964) A suggested origin of conical graupel. J. Appl. Met. 3, 633. [Chapter 4]

HUMPHREYS, W. J. (1940) Physics of the Air, p. 359. McGraw-Hill, New York. [Chapter 7]

INMAN, R. L. (1960) Analysis of AN/CPS-9 radar observations of convective storms. Paper presented at Conference on Severe Storms, St. Louis, Missouri. [Chapter 6]

__________and ARNOLD, J. E. (1961) Thunderstorm characteristics.

Chapter II - Utilization of AN/CPS-9 Radar in Weather Analysis and Forecasting, pp. 8-73. Final Report - Contract AF 19 (604)-6136, A. & M. College of Texas [Now Texas A&M University]. [Chapter 10]

IRIBARNE, J. V. (1968) Development of accumulation zones. Proc. Int. Conf. on Cloud Physics, Toronto, p. 350. [Chapter 8]

__________ and GRANDOSO, H. N. (1965) Experiencia de Modificacion Artificial de Granizadas en Mendoza. Universidad de Buenos Aires, Facultad de Ciencias Exactas y Naturales, Contribuciones Centificas, Serie Meteorologia, Volumen 1, Numero 5. [Chapter 11]

JONES, L. A. (1969) Crop-hail insurance, 1967-68: volume, cost indemnities. U. S. Dept. of Agriculture ERS-424, Washington, D.C. [Chapter 2]

JOUZEL, L., MERLIVAT, L., and ROTH, E. (1975) Isotopic Study of Hail, paper submitted for publication. [Chapter 9]

KARTSIVADZE, A. I. (1967) Experiments in modifying hail phenomena in the Alazani Valley. Trudy Inst. Geofiz. No. 1 (25), Tbilisi, 190. [Chapter 11]

__________ (1968) Modification of hail processes. Proc. Int. Conf. on Cloud Physics, Toronto, p. 778. [Chapter 11]

KIDDER, R. E. and CARTE, A. E. (1964) Structures of artificial hailstones. J. Rech. Atmos. 1, 196. [Chapter 4]

KNIGHT, C. A. and KNIGHT, N. C. (1968) Spongy hailstone growth criteria. I. Orientation Fabrics. J. Atmos. Sci. 25, 445. [Chapter 5]

__________ __________(1970a) Hailstone embryos. J. Atmos. Sci. 27, 659. [Chapter 5]

__________ __________(1970b) Lobe structures of hailstones. J. Atmos. Sci. 27, 667. [Chapter 5]

__________ __________(1971) Hailstones. Scientific American 224, 97. [Chapter 4, 5]

__________ __________(1973) Conical graupel. J. Atmos. Sci. 30, 118. [Chapter 4]

KOENIG, L. R. (1962) A note on a method to determine the orientation of crystals within hailstones. Z. Angew. Math. Phys. (ZAMP) 13, 165. [Chapter 5]

LANGMUIR, I. and BLODGETT, K. B. (1946) A mathematical investigation of water droplet trajectories. U. S. Army Air Forces Tech. Rep. No. 5418. [Appendix A]

LEMONS, H. (1942) Hail in high and low altitudes. Bull. Am. Met. Soc. 23, 61. [Chapter 2]

LETZMANN, L. (1930) Cumulus pulsationen. Met. Z. 47, 236. [Chapter 7]

LEVI, L. and AUFDERMAUR, A. N. (1970) Crystallographic orientation and crystal size in cylindrical accretions of ice. J. Atmos. Sci. 27, 443. [Chapter 5, 8]

LIST, R. (1958) Kennzeichen atmosphärischer Eispartikeln. 1, Teil. Graupeln als Wachstumszentren von Hagelkörnen. Z. angew. Math. Phys. (ZAMP) 9A, 180. [Chapter 4, 5]

__________(1959a) Zur Aerodynamik von Hagelkörnen. Z. angew. Math. Phys. (ZAMP) 10, 143. [Chapter 4, 8]

__________(1959b) Wachstum von Eis-Wassergemischen im Hagelversuchskanal. Helv. Phys. Acta 32, 293. [Chapter 5]

__________(1960a) Design and operation of the Swiss Hail Tunnel. Physics of Precipitation, Geophys. Monogr. No. 5, p. 310. American Geophysical Union, Washington, D.C. [Chapter 5]

__________(1960b) Growth and structure of graupel and hailstones. Physics of Precipitation, Geophys. Monogr. No. 5, p. 317. American Geophysical Union, Washington, D.C. [Chapter 5]

__________(1961a) On the growth of hailstones. Nubila 4, 29. [Chapter 5, 7, 8]

__________(1961b) Physical methods and instruments for characterizing hailstones. Bull. Am. Met. Soc. 42, 452. [Chapter 5]

__________(1963a) General heat and mass exchange of spherical

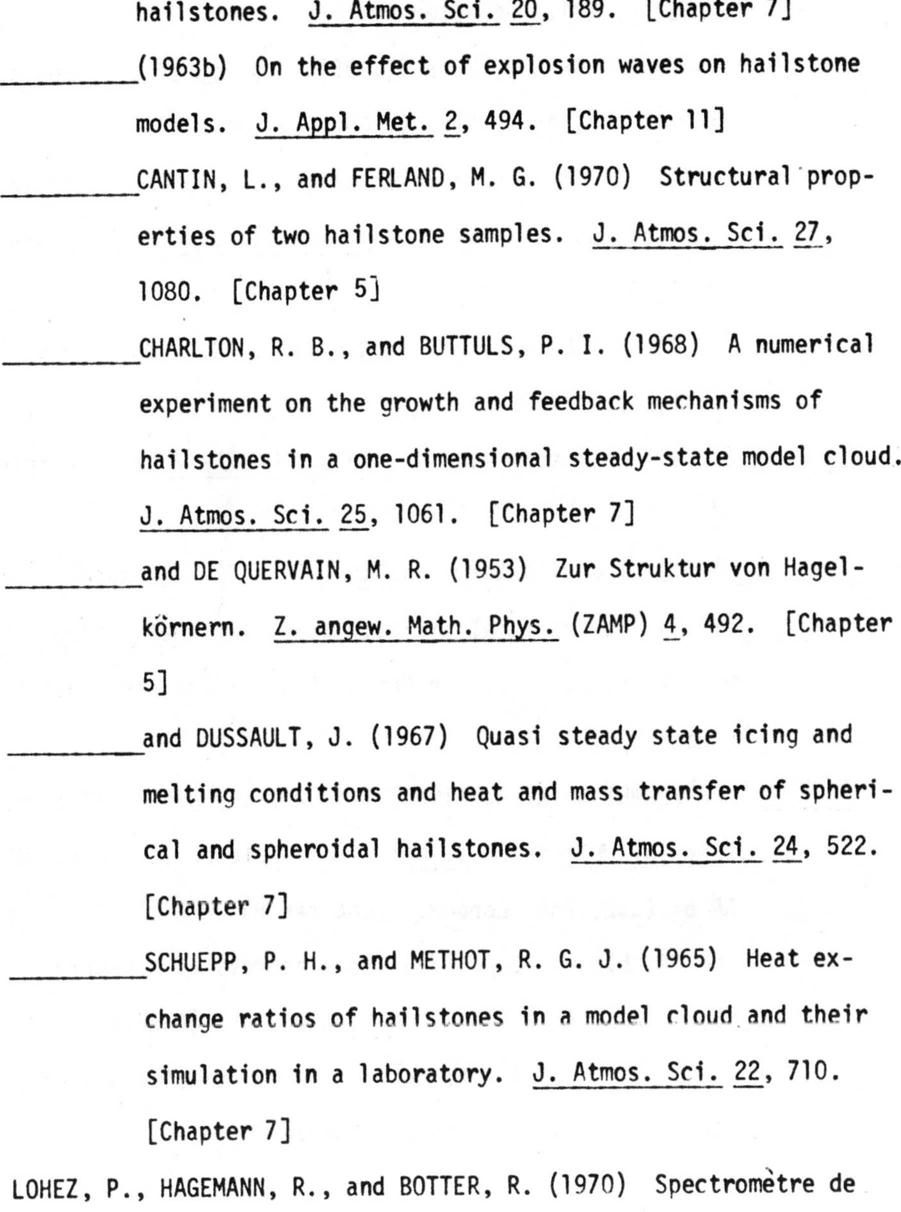

hailstones. J. Atmos. Sci. 20, 189. [Chapter 7]

__________(1963b) On the effect of explosion waves on hailstone models. J. Appl. Met. 2, 494. [Chapter 11]

__________CANTIN, L., and FERLAND, M. G. (1970) Structural properties of two hailstone samples. J. Atmos. Sci. 27, 1080. [Chapter 5]

__________CHARLTON, R. B., and BUTTULS, P. I. (1968) A numerical experiment on the growth and feedback mechanisms of hailstones in a one-dimensional steady-state model cloud. J. Atmos. Sci. 25, 1061. [Chapter 7]

__________and DE QUERVAIN, M. R. (1953) Zur Struktur von Hagelkörnern. Z. angew. Math. Phys. (ZAMP) 4, 492. [Chapter 5]

__________and DUSSAULT, J. (1967) Quasi steady state icing and melting conditions and heat and mass transfer of spherical and spheroidal hailstones. J. Atmos. Sci. 24, 522. [Chapter 7]

__________SCHUEPP, P. H., and METHOT, R. G. J. (1965) Heat exchange ratios of hailstones in a model cloud and their simulation in a laboratory. J. Atmos. Sci. 22, 710. [Chapter 7]

LOHEZ, P., HAGEMANN, R., and BOTTER, R. (1970) Spectromètre de masse automatique pour l'analyse isotopique de l'hydrogène des eaux naturelles. Méthodes Physiques D'Analyse

(GAMS) 6, 291. [Chapter 9]

LONGLEY, R. W. and THOMPSON, C. E. (1965) A study of the causes of hail. J. Appl. Met. 4, 69. [Chapter 10]

LUDLAM, F. H. (1950) The composition of coagulation elements in cumulonimbus. Quart. J. Roy. Met. Soc. 76, 52. [Chapter 7, 8]

__________(1958) The hail problem. Nubila 1, 12. [Chapter 4, 7, 8, 11, Appendix B]

__________(1959) Hailstorm studies, 1958. Nubila 2, 7. [Chapter 6]

__________(1961) The hailstorm. Weather 16, 152. [Chapter 7, 8]

__________(1963) Severe local storms: a review. Met. Monogr. 5, No. 27, p. 1. American Meteorological Society, Boston. [Chapter 6, 7, 9]

__________and BROWNING, K. A. (1960) Radar study of a hailstorm. Imperial College Dept. Met. Tech. Note No. 5, Contract AF 61 (052)-254, London. [Chapter 8]

__________and MACKLIN, W. C. (1959) Some aspects of a severe storm in S. E. England. Nubila 2, 38. [Chapter 2, 3]

__________ __________(1960) The Horsham hailstorm of 5 September 1958. Met. Mag. 89, 245. [Chapter 2]

MACKLIN, W. C. (1961) Accretion in mixed clouds. Quart. J. Roy. Met. Soc. 87, 413. [Chapter 7]

__________(1963) Heat transfer from hailstones. Quart. J. Roy. Met. Soc. 89, 360. [Chapter 7, Appendix B]

__________and BAILEY, I. H. (1966) On the critical liquid water concentrations of large hailstones. Quart. J. Roy. Met. Soc. 92, 297. [Chapter 5, Appendix A]

__________ __________(1968) The collection efficiencies of hailstones. Quart. J. Roy. Met. Soc. 94, 393. [Appendix A]

__________and LUDLAM, F. H. (1961) The fallspeeds of hailstones. Quart. J. Roy. Met. Soc. 87, 72. [Chapter 4, 8]

__________MERLIVAT, L., and STEVENSON, C. M. (1970) The analysis of a hailstone. Quart. J. Roy. Met. Soc. 96, 472. [Chapter 9]

__________and PAYNE, G. S. (1969) The spreading of accreted droplets. Quart. J. Roy. Met. Soc. 95, 724. [Chapter 5]

__________and RYAN, B. F. (1962) On the formation of spongy ice. Quart. J. Roy. Met. Soc. 88, 548. [Chapter 5]

__________RYE, P. J. (1974) Crystallographic orientation distributions in accreted ice. J. Atmos. Sci. 31, 849. [Chapter 5]

__________STRAUCH, W., and LUDLAM, F. H. (1960) The density of hailstones collected from a summer storm. Nubila 3, 12. [Chapter 4, 5]

MAGONO, C. and NAKAMURA, T. (1965) Aerodynamic studies of falling snowflakes. J. Met. Soc. Japan 43, 139. [Chapter 4]

MAJZOUB, M., NIEF, G., and ROTH, E. (1968) Variations and comparisons of deuterium and oxygen 18 concentrations in hailstones. Proc. Int. Conf. on Cloud Physics, Toronto, p. 450. [Chapter 9]

MARSHALL, J. S. (1961) Interrelation of the fall speed of rain and the updraft in hail formation. Nubila 4, 59. [Chapter 8]

_________ and GORDON, W. E. (1957) Radiometeorology. Met. Monogr. 3, No. 14, p. 73. American Meteorological Society, Boston. [Chapter 6, 7]

MARWITZ, J. D. (1971) Supercell storms and the Soviet hailstorm model. Proc. Int. Conf. on Weather Modification, Canberra, p. 207. [Chapter 8]

_________ (1972a) The structure and motion of severe hailstorms. Part III: Severely Sheared Storms. J. Appl. Met. 11, 189. [Chapter 7]

_________ (1972b) The structure and motion of severe hailstorms. Part I: Supercell Storms. J. Appl. Met. 11, 166. [Chapter 7]

_________ (1973) Hailstorms and hail suppression techniques in the U.S.S.R. - 1972. Bull. Am. Met. Soc. 54, 317. [Chapter 11]

MC BRIDE, J. H. (1964) Small-scale structure of hail swaths. M.Sc. thesis, Dept. of Meteorology, McGill University.

(Unpublished manuscript) [Chapter 3]

MERLIVAT, L. and NIEF, G. (1967) Fractionnement isotopique lors des changements d'état solide-vapeur et liquid-vapeur de l'eau à des températures inférieures à 0°C. Tellus 19, 122. [Chapter 9]

__________ __________and ROTH, E. (1965) Formation de la grêle et fractionnement isotopique du derterium. Abhandlungen der deutschen Akademie der Wissenschaften zu Berlin 7, 839. [Chapter 9]

MÖLLER, M. (1884) - Met. Z. 1, 230. [Chapter 7]

MORGAN, G. M., Jr. (1970) An examination of the wet-bulb zero as a hail forecasting parameter in the Po Valley, Italy. J. Appl. Met. 9, 537. [Chapter 10]

__________(1973) A general description of the hail problem in the Po Valley of Northern Italy. J. Appl. Met. 12, 338. [Chapter 2, 11]

MOSSOP, S. C. (1971) Some hailstones of unusual shape. Weather 26, 222. [Chapter 4]

__________and KIDDER, R. E. (1961) Hailstorm at Johannesburg on 9th November 1959: Part II - Structure of hailstones. Nubila 4, 74. [Chapter 4, 5]

MÜLLER, H. G. (1964) I. Bericht über die Hagelabwehrvesuche im Landkreis Rosenheim. Deutsche Versuchsanstalt für Luft- und Raumfahrt e. V. [Chapter 11]

__________(1967) Weather modification experiments in Bavaria. _Proc. 5th Berkeley Symposium on Mathematical Statistics and Probability_, Vol. 5, p. 223. University of California Press, Berkeley. [Chapter 11]

MUSIL, J. D. (1970) Computer modeling of hailstone growth in feeder clouds. _J. Atmos. Sci._ _27_, 474. [Chapter 7]

NAKAYA, U. (1954) _Snow Crystals: Natural and Artificial._ Harvard University Press, Cambridge. [Chapter 4]

NATIONAL ACADEMY OF SCIENCES - NATIONAL RESEARCH COUNCIL (1966) _Weather and Climate Modification - Problems and Prospects - VOLUME I: Summary and Recommendations, VOLUME II: Research and Development_. Final Report of the Panel on Weather and Climate Modification to the Committee on Atmospheric Sciences, National Academy of Sciences, National Research Council. Publication No. 1350, Washington, D.C. [Chapter 11]

NATIONAL CENTER FOR ATMOSPHERIC RESEARCH (1969) _Plan for the Northeast Colorado Hail Experiment_. [Prepared by NCAR in consultation with the Select Planning Group of the Northeast Colorado Hail Experiment - Boulder, Colorado - 17 March 1969.] [Appendix C]

__________(1972) _The National Hail Research Experiment - Summer 1972 Summary Report_. NCAR, Boulder, Colorado. [Chapter

8, Appendix C]

__________(1973) The National Hail Research Experiment - Summer 1973 Summary Report. NCAR, Boulder, Colorado. [Appendix C]

__________(1974) The National Hail Research Experiment - Summer 1974 Summary Report. NCAR, Boulder, Colorado. [Appendix C]

__________(1974) The National Hail Research Experiment - Project plan for 1975 to 1980. NCAR, Boulder, Colorado. [Appendix C]

NEWMAN, S. and GOKHALE, N. R. (1975) The melting rates of various size hailstones in a vertical wind tunnel. (Unpublished manuscript) [Appendix B]

NEWTON, C. W. (1963) Dynamics of severe convective storms. Met. Monogr. 5, No. 27, p. 33. American Meteorological Society, Boston. [Chapter 7, 8]

__________and NEWTON, H. R. (1959) Dynamical interactions between large convective clouds and environment with vertical shear. J. Met. 16, 483. [Chapter 8]

PAINTER, P. R. and SCHAEFER, V. J. (1960) Permanent replicas of the crystalline structure of hailstones. Z. angew. Math. Phys. (ZAMP) 11, 318. [Chapter 5]

PAPPAS, J. J. (1962) A simple yes-no hail forecasting technique.

J. Appl. Met. 1, 353. [Chapter 10]

PAUL, A. H. (1968) Regional variations in two fundamental properties of hailfalls. Weather 23, 424. [Chapter 2]

PELL, J. (1969) The Alberta hailstorm as observed on the ground and by radar. M.Sc. thesis, McGill University. (Unpublished manuscript) [Chapter 3]

PETTERSSEN, S. (1969) Introduction to Meteorology. Third edition, McGraw-Hill Book Co., New York. [Chapter 2, 7]

PHILLIPS, B. B. (1969) Water load in convective storms and its influence on storm kinetics. ESSA Technical Memorandum ERLTM-APCL 6, U. S. Dept. of Commerce, Washington, D.C. [Chapter 8]

POWELL, G. (1961) The relationship of physiography to the hail distribution pattern in central and southern Alberta. M.Sc. thesis, Dept. of Geography, University of Alberta. (Unpublished manuscript) [Chapter 2]

PROCTOR, D. and GEOTIS, S. (1963) Ground patterns of hailstorms. Proc. 10th Weather Radar Conf., Washington, D.C., p. 72. [Chapter 6]

PROHASKA, K. (1900) Die jahrliche und tagliche Periode der Gewitter und Hagelfalle in Steiermark und Kaernten. Met. Z. 17, 327. [Chapter 3]

__________(1907) Die Hagelfalle des 6. Juli 1905 in den Ostalpen. Met. Z. 24, 193. [Chapter 3]

RAMDAS, L. A., SATAKOPAN, V., and RAO, S. G. (1938) Frequency of days with hailstorms in India. India J. Agri. Sci. 8, 787. [Chapter 2]

RANZ, W. E. and MARSHALL, W. R. (1952) Evaporation from drops, Part I and II. Chem. Eng. Prog. 48, 141, 173. [Chapter 7, Appendix B]

RATNER, B. (1961) Do high-speed winds aloft influence the occurrence of hail? Bull. Am. Met. Soc. 42, 443. [Chapter 7, 8]

REYNOLDS, D. (1876) Formation of raindrops and hailstones. Nature, 15, 163. [Chapter 4]

ROGERS, L. N. (1971) Two unusual hailstones. Bull. Am. Met. Soc. 52, 994. [Chapter 5]

ROOS, D. V. D. S. (1972) A giant hailstone from Kansas in free fall. J. Appl. Met. 11, 1008. [Chapter 4]

__________ CARTE, A. E. (1973) The falling behavior of oblate and spiky hailstones. J. Rech. Atmos., VII, 39.[Chapter 4]

SÄNGER, R. (1960) The mechanism of hail formation. Physics of Precipitation, Geophys. Monogr. No. 5, 305. American Geophysical Union, Washington, D.C. [Chapter 11]

__________ SPRING, F., STAUB, W., and THAMES, J. C. (1962) Grossversuch III zur bekampfung des hagels im tessin, Tatigkeitsbericht 14. Eidg. Kommission, Studium d.

Hagelbildung und der Hagelabwehr, Bern, 102. [Chapter 11]

SANSOM, H. W. (1965) A preliminary report on a hail suppression experiment in Kenya. Proc. Int. Conf. on Cloud Physics, Tokyo and Sapporo, 449. [Chapter 11]

__________(1966) The occurrence and distribution of hail in Africa. Met. Mag. 95, 212. [Chapter 2]

SARRICA, O. (1965) Observational results on hail formation and structure. Ricerca Scient. 35, 345. [Chapter 5]

SCHAEFER, J. J. (1960) Hailstorms and hailstones of the Western Great Plains. Nubila 4, 18. [Chapter 5]

SCHLEUSENER, R. A. (1962) The 1959 hail suppression effort in Colorado and evidence of its effectiveness. Nubila 5, 31. [Chapter 11]

__________(1966) Project Hailswath: Final Report, Volume 2, Reports from participating groups. Rapid City, South Dakota. [Chapter 3]

__________(1968) Hailfall damage suppression by cloud seeding - a review of the evidence. J. Appl. Met. 7, 1004. [Chapter 11]

__________and AUER, A. H. (1964) Hailstorms in the High Plains - Final Report. Civil Engineering Section, Colorado State University, Fort Collins, Colorado. [Chapter 3]

SCHMID, P. (1967) On Grossversuch III, a randomized hail

suppression experiment in Switzerland. Proc. 5th Berkeley Symposium on Mathematical Statistics and Probability, Vol 5, p. 141. University of California Press, Berkeley. [Chapter 11]

SCHUMANN, T. W. W. (1938) The theory of hailstone formation. Quart. J. Roy. Met. Soc. 64, 3. [Chapter 4, 7]

SHOWALTER, A. (1953) A stability index for thunderstorm forecasting. Bull. Am. Met. Soc. 34, 250. [Chapter 10]

SIMPSON, G. C. (1925) Thunderstorms and aviation. J. Roy. Aeronaut. Soc. 29, 24. [Chapter 7]

SIMPSON, J. and WIGGERT, V. (1969) Models of precipitating cumulus towers. Mon. Wea. Rev. 97, 471. [Chapter 8]

SMITHSONIAN INSTITUTION (1951) Smithsonian Meteorological Tables. Washington, D.C. [Appendix D]

SPENGLER, J. D. (1971a) Experimental studies of hydrometeor interactions using a large vertical wind tunnel. Ph.D. thesis, Dept. of Atmospheric Science, State University of New York at Albany. (Unpublished manuscript) [Chapter 5]

__________ (1971b) Eleventh Yellowstone Field Research Expedition - Final Report, p. 96. Atmospheric Sciences Research Center, State University of New York at Albany, New York. [ASRC-SUNYA Publication #141] [Chapter 5]

__________ and GOKHALE, N. R. (1970) Large vertical wind tunnel

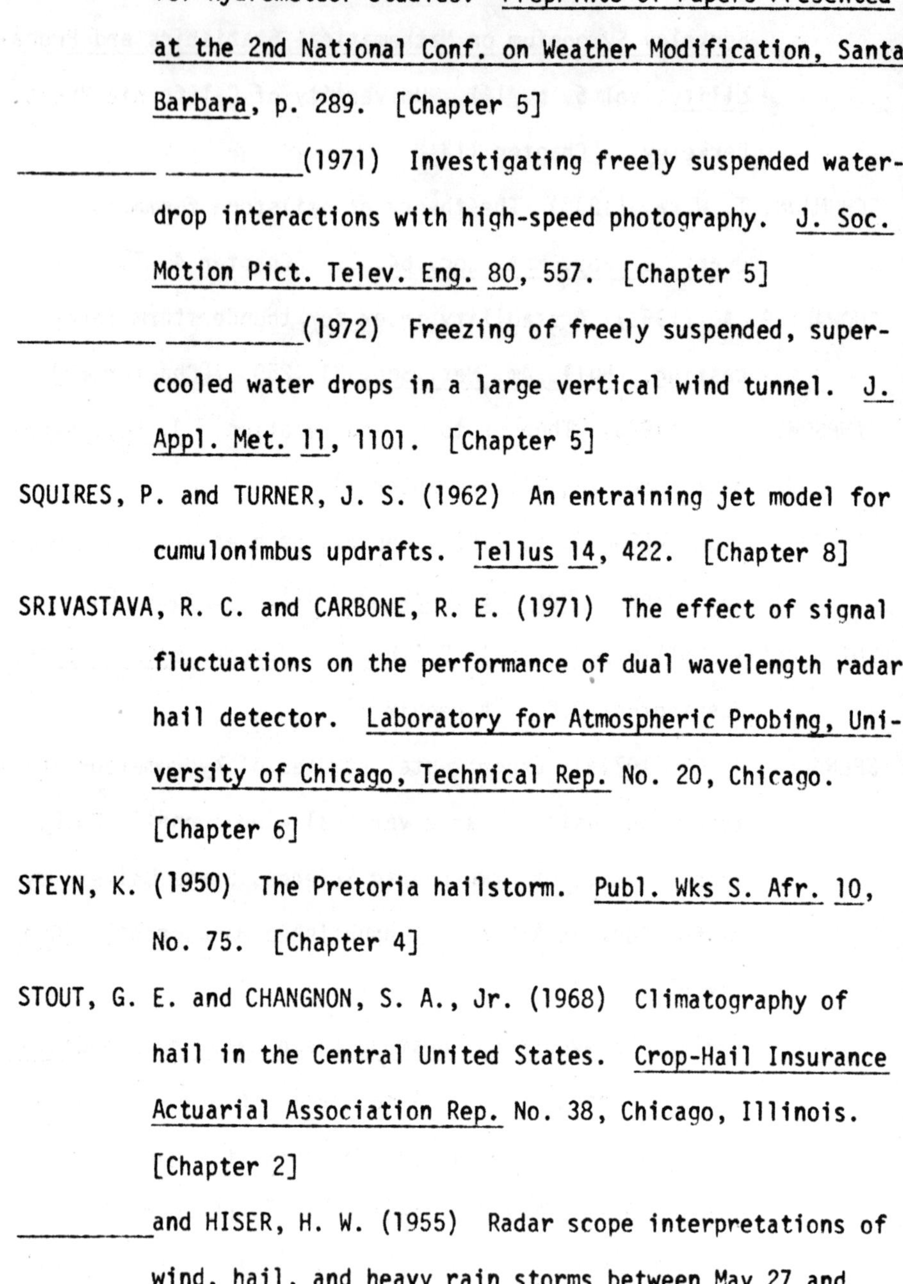

for hydrometeor studies. Preprints of Papers Presented at the 2nd National Conf. on Weather Modification, Santa Barbara, p. 289. [Chapter 5]

__________ __________(1971) Investigating freely suspended water-drop interactions with high-speed photography. J. Soc. Motion Pict. Telev. Eng. 80, 557. [Chapter 5]

__________ __________(1972) Freezing of freely suspended, super-cooled water drops in a large vertical wind tunnel. J. Appl. Met. 11, 1101. [Chapter 5]

SQUIRES, P. and TURNER, J. S. (1962) An entraining jet model for cumulonimbus updrafts. Tellus 14, 422. [Chapter 8]

SRIVASTAVA, R. C. and CARBONE, R. E. (1971) The effect of signal fluctuations on the performance of dual wavelength radar hail detector. Laboratory for Atmospheric Probing, University of Chicago, Technical Rep. No. 20, Chicago. [Chapter 6]

STEYN, K. (1950) The Pretoria hailstorm. Publ. Wks S. Afr. 10, No. 75. [Chapter 4]

STOUT, G. E. and CHANGNON, S. A., Jr. (1968) Climatography of hail in the Central United States. Crop-Hail Insurance Actuarial Association Rep. No. 38, Chicago, Illinois. [Chapter 2]

__________and HISER, H. W. (1955) Radar scope interpretations of wind, hail, and heavy rain storms between May 27 and

June 8, 1954. Bull. Am. Met. Soc. 36, 519. [Chapter 6]

SUCKSTORFF, G. A. (1939) Die Ergebnisse der Untersuchungen an tropischen Gewittern. Gerlands Beitr. Geophys. 55, 138. [Chapter 7]

SULAKVELIDZE, G. K. (1966a) Findings of the Caucasus Anti-Hail Expedition (1965). (Translated from Russian by Israel Program for Scientific Translations, Jerusalem. 1967) National Science Foundation, Washington, D.C. [Chapter 6, 11]

__________(1966b) Principles and means of modifying convective clouds for the purpose of hail prevention used during the 1964-1965 Caucasus expedition. Trudy Vysok. Geofiz. Inst. No. 5, Leningrad, 150. [Chapter 11]

__________(1967) Rainstorms and Hail. (Translated from Russian by Israel Program for Scientific Translation, Jerusalem. 1969) National Science Foundation, Washington, D.C. [Chapter 2, 3, 7, 8]

__________(1968) On the principles of hail control method applying in the USSR. Proc. Int. Conf. on Cloud Physics, Toronto, p. 796. [Chapter 11]

__________BIBILASHVILI, N. Sh., and LAPCHEVA, V. F. (1965) Formation of Precipitation and Modification of Hail Processes. (Translated from Russian by Israel Program for

Scientific Translations, Jerusalem. 1967) National Science Foundation, Washington, D.C. [Chapter 7, 8, 11]

SUMMERS, P. W. (1968) Soft hail in Alberta hailstorms. Proc. Int. Conf. on Cloud Physics, Toronto, p. 455. [Chapter 4]

__________ and PAUL, A. H. (1967) Some climatological characteristics of hailfall in central Alberta. ALBERTA HAIL STUDIES 1967, McGill Univ. Stormy Weather Group Sci. Rep. MW-57, p. 17, Montreal. [Chapter 2]

__________ and WOJTIW, L. (1971) The economic impact of hail damage in Alberta, Canada and its dependence on various hailfall parameters. Preprints of Papers Presented at the 7th Conf. on Severe Local Storms, Kansas City, p. 158. [Chapter 2]

__________ MATHER, G. K. and TREDDENICK, D. S. (1972) The development and testing of an airborne droppable pyrotechnic flare system for seeding Alberta hailstorms. J. Appl. Met. 11, 695. [Chapter 11]

THORPE, A. D. and MASON, B. J. (1966) The evaporation of ice spheres and ice crystals. Brit. J. Appl. Phys. 17, 541. [Chapter 7]

TOWERY, N. G., STAGGS, D. W., and LONNQUIST, C. G. (1970) Hailstorm characteristics as viewed by Rapid Scan RHI Radar.

Preprints of Papers Presented at the 2nd National Conf. on Weather Modification, Santa Barbara, p. 172. [Chapter 6]

U. S. WEATHER BUREAU (1947) Thunderstorm rainfall. U. S. Weather Bureau Hydromet. Rep. No. 5, Washington, D. C. [Chapter 2]

VISHER, S. S. (1966) Climatic Atlas of the United States. Harvard University Press, Cambridge. [Chapter 2]

VITTORI, O. (1960) Preliminary note on the effect of pressure waves upon hailstones. Nubila 3, 34. [Chapter 4, 11]

__________ and DI CAPORIACCO, G. (1959) The density of hailstones. Nubila 2, 51. [Chapter 4]

VORONOV, G. S., GAYVORONSKII, I. I., LESKOV, B. N., and SEREGIN, Yu. A. (1967) Experimental hail protection in the Moldavian SSR. Meteor. i. Gidrol. No. 7, Moscow, 29. [Chapter 11]

WEGENER, A. (1911) Thermodynamik der Atmosphare. J. A. Barth, Leipzig. [Chapter 7]

WEICKMANN, H. K. (1953) Observational data on the formation of precipitation in cumulonimbus clouds. In Thunderstorm Electricity (ed. H. R. Byers), p. 66. University of Chicago Press, Chicago. [Chapter 3, 4, 5, 7, 11]

__________(1957) Physics of precipitation. Met. Monogr. 3, No. 19, p. 226. American Meteorological Society, Boston. [Chapter 7]

__________(1964) The language of hailstorms and hailstones. Nubila 6, 7. [Chapter 2, 8, 11]

__________(1969) ESSA - 1968 - Colorado Hail Research. Preprints of Papers Presented at the 6th Conf. on Severe Local Storms, Chicago, p. 314. (Unpublished manuscript) [Chapter 8]

WILK, K. E. (1960) An investigation of a severe local hailstorm. Proc. 8th Weather Radar Conf., San Francisco, p. 481. [Chapter 6]

__________(1961) Radar reflectivity observations of Illinois thunderstorms. Proc. 9th Weather Radar Conf., Kansas City, p. 127. [Chapter 6, 10]

WILLIAMS, G. N. and DOUGLAS, R. H. (1963) Alberta storms. ALBERTA HAIL STUDIES 1962/1963, McGill Univ. Stormy Weather Group Sci. Rep. MW-36, p. 17, Montreal. [Chapter 3]

WILLIS, J. T., BROWNING, K. A., and ATLAS, D. (1964) Radar observations of ice spheres in free fall. J. Atmos. Sci. 21, 103. [Chapter 4]

WINSTON, J. S. (1956) [Editor] Forecasting Tornadoes and Severe Thunderstorms. U. S. Weather Bureau Forecasting Guide

No. 1 [Now National Weather Service]. [Chapter 10]

WISNER, C., ORVILLE, H. D. and MYERS, C. (1972) A numerical model of a hail-bearing cloud. J. Atmos. Sci. 29, 1160. [Chapter 7]

WOOD, C. P. (1955) A theory of hail formation. Master's thesis, Massachusetts Institute of Technology. (Unpublished manuscript) [Chapter 10]

YOUNG, R. G. E. and BROWNING, K. A. (1967) Wind tunnel tests of simulated spherical hailstones with variable surface roughness. J. Atmos. Sci. 24, 58. [Chapter 4]

SUBJECT INDEX